Q P

Costas Tsiamis
Plague in Byzantine Times

Medicine in the Medieval Mediterranean

Edited by
Alain Touwaide

Scientific Committee:
Vivian Nutton, Marie Hélène Congourdeau,
Dimitri Gutas and Filippo Ronconi

Volume 9

Costas Tsiamis

Plague in Byzantine Times

A Medico-historical Study

DE GRUYTER

ISBN 978-3-11-061119-9
e-ISBN (PDF) 978-3-11-061363-6
e-ISBN (EPUB) 978-3-11-061125-0
ISSN 2569-314X

Library of Congress Control Number: 2022945601

Bibliographic information published by the Deutsche Nationalbibliothek
The Deutsche Nationalbibliothek lists this publication in the Deutsche Nationalbibliografie;
detailed bibliographic data are available on the Internet at http://dnb.dnb.de.

www.degruyter.com

To Chryssa

There is nothing more deceptive
than an obvious fact...
Sir Arthur Conan Doyle

Foreword
Ancient Epidemics
The Necessity of a Multidisciplinary Approach

In this well-researched, insightful monograph *Plague in Byzantine Times: A Medico-historical Study*, Costas Tsiamis accepts the challenge of a topic that has long right-fully been considered a *vexata quaestio* in the history of both medical practice and pathology: ancient epidemics. To this date it has proved particularly complicated for scholars to determine the exact aetiology of infectious diseases that caused epi-demico-pandemic manifestations, as well as their epidemiology, clinical presenta-tions, and disappearance. A classic example of this, also aptly discussed in this study, is the Plague of Athens that devastated the ancient Attic *polis* in the 5th century B.C. In combination with the Peloponnesian War, plague provoked the decline of Athens and represented its farewell to its long-held imperial ambitions.

Despite undeniable advances in the research methodologies and the biomedical techniques that have opened a new era for medico-historical discoveries, ancient epi-demics and pandemics are still more of a conundrum than a *terra cognita* to be crossed at ease. Questions and doubts are more numerous than answers. This is es-pecially true if one considers not only the antiquity of diseases, but also their evolu-tionary history. One may be too hastily tempted to believe that the clinical presenta-tion of diseases has indeed always been the same throughout the centuries, whereas it cannot be excluded that their semiology may have changed owing to pathomor-phosis, which could reflect genetic mutations in the pathogens and their interaction with their hosts. This is something that ought not to be neglected and should be con-sidered when examining ancient literary accounts of epidemics.

Moreover, as it emerges from Tsiamis' pages, the ancient literary sources should be approached with extreme caution as they may use a language and knowledge that predate, sometimes by many a century, the discoveries of contemporary medicine. The very terms used to describe the clinical manifestations of nosological conditions should be carefully analyzed before any conclusions, either aiming to be definitive or being preliminary, can be drawn and put forward, and multiple sources should be scrutinized in the search for evidence. Such evidence could be made of data extract-ed not only from literary accounts, but also from archaeological excavations, numis-matics, and ancient human remains, among others.

In addition, possible interpretative confusions and the inherent limitations of the matter subjected to the study of scholars devoting their academic activity to these topics, are not only an issue for the assessment of the past of infectious diseases, but also for the present, as the COVID-19 pandemic has recently demonstrated. This is particularly true when one reflects on the general despair encountered in both world leadership and populations facing a new, so-called "invisible enemy", which may appear a novel phenomenon at first glance, whereas, at closer look

https://doi.org/10.1515/9783110613636-001

and availing oneself of the tools of historico-medical research, it can be shown to have always existed and have characterised humankind's response to such epochal manifestations of unexplained forces, alternatively interpreted as the results of divine wrath or some conspiracy ascribed to certain ethnic or religious minorities plotting against constituted authority. Furthermore, in general, it can be stated that epidemics mark epoch transitions or even end up accelerating them, something that makes their study of peculiar transdisciplinary interest.

Examining Justinian's interaction with the political and religious institutions in a time of epidemics in the Byzantine Empire is relevant to the understanding of the complex socio-cultural dynamics taking place in the current pandemic times. Let us consider, for instance, the law Justinian issued on March 23, 543 A. D. decreeing that God's teaching (i. e., the punishment of Humankind through the plague) was over. That decision had social and economic reasons, in spite of the fact that the plague would still continue to exist in the Empire and would continue to kill a high number of its subjects: this can be catalogued as a form of an *ante litteram* manipulation of the epidemiological reality, something that many a government worldwide attempted to implement in recent times when protests against restrictions of social interactions or economic hardships emerged in the COVID-19 pandemic.

Distancing himself from those approaches that value one method over the other, Costas Tsiamis rightly chooses to look at the complex matter lying before his eyes, adopting a multidisciplinary methodology that combines historical, archaeological, anthropological, and palaeopathological approaches in a very appropriate blend.

While much is still to be written about the nature and presentations of ancient epidemics, including those from the Byzantine Empire that Costas Tsiamis dwells a great deal upon in his research, *Plague in Byzantine Times* shows how the past of a disease can be fully understood only if methods are combined and knowledge is shared between scholars from various fields of research, never forgetting that Science is one, while its interpreters and disciples may be multiple.

Francesco M. Galassi
Flinders University
June 2022

Contents

Acknowledgments

The present book is a revision of my doctoral thesis *Historical and Epidemiological Approach of Plague during the Byzantine Times (330 – 1453 A.D.)*, which I wrote between 2005 and 2010 at the Department of Hygiene, Epidemiology and Medical Statistics of the Medical School of the National and Kapodistrian University of Athens. I would like to express my sincere gratitude to those who guided and trusted me in that time, namely my tutors Professor Eleni Petridou (Epidemiology), Professor Effie Poulakou-Rebelakou (History of Medicine), and Professor Aristotelis Efthychiades (History of Medicine). While delving into the historical sources over many years, I was constantly accompanied by Professor Dionysios Stathakopoulos (Department of History and Archaeology, University of Cyprus) and Dr. Ioannis Telelis (Academy of Athens, Research Center for Greek and Latin Literature), to whom I am most grateful for offering their specialized knowledge whenever I needed it.

The molecular study of the plague pathogen rapidly developed after I completed my doctoral thesis, making it necessary to correct my work, to add new elements, and to update data. In the past years I have had the great privilege to do research at the Department of Microbiology of the Medical School of the National and Kapodistrian University of Athens, where I found myself right at the heart of the laboratory diagnosis of infectious diseases and became accustomed with the most modern diagnostic methods. I wish to thank the Director of the Department of Microbiology, Professor Athanassios Tsakris, as well as Professor Georgia Vrioni, for the knowledge they shared with me and the help they provided to penetrate the secrets of the microcosm and approach *Yersinia pestis*.

The epidemiological approach to an infectious disease comprises two basic parameters: space and time. On these matters, I wish to thank Professor Dimitrios Anoyatis-Pelé (School of History, Ionian University, Corfu) for his help on issues related to Historical Demography and Geography, and Alexandros Aidonidis from the Historical Demography Postgraduate Program of the Ionian University, who edited the plague maps that frame and illustrate the history of the disease.

Paraphrasing Captain Cook, my childhood hero who considered the first navigation clock to be his 'most faithful guide that shows the right path', my friend and long-term colleague and translator Aggelos Zikos, constantly showed me the right path in the uncharted waters of my endeavour over the past years. I deeply thank him for the translations and his technical skills that helped me to make my writing more understandable.

The publication of this book would not have been possible without the assistance, advice and trust placed in me by the Senior Editor of the series *Medicine in the Medieval Mediterranean*, Dr. Alain Touwaide (Scientific Director, Institute for the Preservation of Medical Traditions, Washington, D.C., USA), to whom I am deeply grateful. I also wish to warmly thank the other members of the editorial board: Vivian Nutton (Emeritus Professor of History of Medicine at University College London),

https://doi.org/10.1515/9783110613636-002

Marie Hélène Congourdeau (Honorary Researcher at the Centre National de la Recherche Scientifique, Paris-France) and Dimitri Gutas (Professor of Arabic and Graeco-Arabic Studies at Yale University).

Also, I wish to thank Dssa Emanuela Appetiti (President, Institute for the Preservation of Medical Traditions, Washington, D.C., USA) for her enthusiastic support and her detailed proofreading of the final version of my manuscript. In this final phase Marie Hélène Congourdeau, Niki Papavramidou and Witold Witakowski have been helpful in checking documents difficult to trace in a time when library are still close because of the pandemic. As usual all mistakes remain mine.

Finally, I warmly thank De Gruyter Senior Editor for Classical Studies, Dr Serena Pirrotta, and her team, especially Marco Acquafredda and Dr Carlo Vessella, as well as all De Gruyter dedicated staff for their collaboration and patience, professionalism, and assistance with technical issues.

Costas Tsiamis
Athens, June 2022

Introduction

Let us take a brief moment to consider whether the human brain—given the amount of information and knowledge it stores—is able to establish a spontaneously logical link upon hearing the following words: Lake Victoria, Damascus, hibernation. An instinctive and probably the sole, reasonable answer would be that no connection exists between them. When the brain visualizes these words, it might automatically visualize three different pictures, and three distinctive and distant geographical environments and ecosystems: savanna, desert, and steppe, respectively. What if these images, however, were not only related to space but also to time? Of course, the concept of time is 'flexible' and may refer to the alternation of seasons during the calendar year, the alternation of aeons in the desert, or to a given point in time in the steppe. What would happen though, when the projections of events that took place at different times and places start to intersect at a specific point in time?

This study attempts to broaden the way we think and approach the biological phenomenon of epidemics. This is the first inherent obstacle to this effort. The exchange of information between different scientific fields is given. What kind of information is needed, however, and by whom? Diffusion and access to information is easy nowadays. How easy is it for a researcher, however, to understand the findings of another scientific field that is different from his/her own? Modern scientific collaborations lead to interesting and useful conclusions which are easily disseminated and exchanged. Supposing that a researcher has access to a plentiful amount of data, what would the key-element allowing him/her to 'unlock' the sequence of events be? It is easy for the reader to understand the difficulties stemming from the approach to the same issue by two distinct sciences, from a different starting point. It is understood that when we examine this issue retrospectively over the course of centuries, its solution becomes exponentially more difficult, the more so when it comes to health-related issues and especially to epidemics.

Studying the epidemics of the distant past poses a particular challenge and, regardless of how theoretical our approach is, such studies present certain scenarios which are more or less likely to be closer to the facts. Research is not one-sided, but rather multidimensional, as it draws data from various sources. This effort can be likened to the pieces of a puzzle. Before fitting the pieces, we need to separate and group them together by colour or contour. Even this is not always enough to find a solution, however. Epidemics are multifactorial systems that entail subtle and often long-term interactions. Each process and factor represents a distinct piece of the broader picture. The aim of this study is to present as many pieces of this historical puzzle as possible. The theories put forward herein are based on this very method. The distinct pieces of the same system that started to come together as a result of historical events, biological processes or even randomly, are all laid out right before us. What remains is to reconstruct the full picture which, as already men-

https://doi.org/10.1515/9783110613636-003

tioned, shall always comprise an element of probability rather than absolute certainty. The reader must nevertheless not forget that the study of epidemics moves along two axes: space and time.

This study aims at collecting, presenting and processing historical and medical data on the plague epidemics that struck the Byzantine Empire. It will present the two plague pandemics, that is, the *Plague of Justinian* and the *Black Death*, which ravaged the Byzantine Empire. Our approach, namely the combination of historical data and contemporary medical knowledge, highlights the importance of interdisciplinary cooperation in the field of Historical Epidemiology, History of Public Health, and the History of Medicine in general. The range of scientific fields involved in this study is wide, thus providing for the use of data from several disciplines such as Epidemiology, Microbiology, History, Historical Demography and Geography, as well as Archaeology.

Nosologically, plague exists until today and every year a few countries report plague cases to the World Health Organization (WHO), which by no means present the intensity and extent that characterized the disease some centuries ago. The medical interest in plague remains high and it is indicative that the keyword *plague* came up in more than 12,000 publications on the *MedLine* database, which cover the period 1960 – 2021. After 2001, plague resurfaced along with smallpox and anthrax, in the context of bio-terrorism after the attacks of September 11 in New York, given that *Yersinia pestis* is listed as a Class A biological weapon. Beyond this politico-military dimension, however, the study of plague continues in the context of analyzing the genome of *Yersinia pestis*, in order to prepare effective vaccines and to predict, prevent, detect, and control the natural foci of the disease all over the world by the WHO and the national health services.

This study will cover the whole spectrum of plague epidemics that occurred in the Byzantine Empire, from the year of its foundation until its fall, that is, from 330 to 1453 A.D. The time of occurrence of the two pandemics coincides with two particular periods of Byzantine history. The first period (541–750) relates to the acme of the Empire, whereas the second period (1347–1453) coincides with the struggle of a state entity that was discredited on the political scene of the fourteenth and fifteenth centuries, and was tormented by civil wars, corruption, government decay, and decadence until its final fall. The epidemics of the first pandemic are presented in the context of a 'Byzantine ecumene', whereas subsequent pandemic comprise the epidemics that occurred in territories under foreign occupation, located mainly in the regions of Greece, thus spatially delimiting a wider geographical area perceived as 'Byzantine space'.

Previous studies have highlighted specific questions which have not yet been dealt with, even more so at a time when epidemiological data are inconsistent with prevailing views. According to one view, for example, the biblical *Plague of the Philistines*, the *Plague of Athens*, the *Antonine Plague* or the *Plague of Cyprian* constitute cases of plague epidemics. On the other hand, despite the indication that these epidemics affected immunologically 'virgin' populations, there is no factual evidence that the disease existed before the fifth century A.D. This view, however, which is often reproduced in medical textbooks of microbiology and epidemiology as part of their introduction to plague, illustrates the tradition of obsolete theories. The questions are not exhausted here, though. The present study attempts to clarify such questions arising from reading the sources. As far as the first pandemic is concerned (the so-called *Plague of Justinian*), the question of the gateway of the disease into Europe and the origin of the first pandemic remains. In an effort to identify the origin of plague, the phenomenon of research orientation (or disorientation) towards identifying epidemics before 541 instead of monitoring the course of the microbe is presented. The present study aspires to combine existing views, taking into account not only historical data, but also particular phylogenetic and evolutionary elements of *Yersinia pestis* strains. In this context, it is deemed necessary to redefine the grouping of epidemic outbreaks. Studies have so far been linking the disease with trade, wars and movements of populations, disregarding the possible existence of enzootic foci.

Similarly, the *Second Pandemic* (or *Black Death*) displays particular elements that need clarification. The medico-historical study is entrapped into the narrative of Gabriele de' Mussi, which covers the events of the terrible siege of the Genoese colony of Kaffa in Crimea by the Mongols in 1346, wherefrom plague spread to Europe, after affecting Constantinople in 1347. The incubation period, the nature of the disease, the time of the siege and particular Byzantine sources, all prompt us to think of other possible routes that plague followed from Crimea to Constantinople. Our study attempts to present the epidemiological and historical data of the period from a different perspective, that is, in conjunction with the findings of the first pandemic, instead of examining only the period in question. In addition, it has been deemed appropriate to compare the clinical forms of epidemics of the two pandemics, given that the medical and medico-historical literature has revealed controversies over the true nature of the *Second Pandemic*. Finally, in the case of *Black Death* epidemics, the grouping of epidemic waves and outbreaks is considered equally necessary.

This study takes into account a large number of works carried out to date. It rests on two fundamental elements defined by the modern medical community: the clinical picture of plague, and its epidemiological cycle. These two elements constitute our interpretive and comparative tool for describing the Byzantine disease. They are essentially critical of primary sources and, by extension, secondary literature since—as already mentioned—it is a well known fact that positions and opinions ultimately need to be revised.

Unfortunately, the era of the two pandemics has been misinterpreted and the general public does not usually grasp that these centuries are crucial to the shaping of the European culture and modern European states. To some extent, the readers' preconception, which is often the result of the study of epidemics of this period, should be attributed to the term 'Middle Ages' which, by association and insufficient knowledge, suggests a *dark period*. In the context of the misunderstood significance of this period, which supposedly slowed down the progress of Europe, epidemics are construed as an expected result of superstition and ignorance of contemporary people, while the social behaviors that emerged seem strange and bizarre nowadays. However, when we examine the epidemics of that period from a purely epidemiological perspective, we may conclude that infectious diseases have diachronically been governed by fixed epidemiological cycles and models and that, as such, they can emerge anytime and anywhere on the planet.

The number of years that separates us from that era is minimal in relation to the actual presence of the *Homo sapiens* on earth for tens of thousands of years. Beyond any cultural differences, modern humans appear to share psychological traits with the people living in the Middle Ages. Collective fears and emotional reactions were, are and will remain, a human characteristic. Moreover, despite the sophisticated diagnostic and therapeutic arsenal of modern Medicine, nobody could guarantee that, in the case of a modern pandemic of any deadly disease, no similar individual and collective social behavior would emerge. A behavior may once again go to extremes, from fleeing and abandonment to pogroms against social, national or religious groups. Therefore, if ignorance was the curse of people of the Middle Ages and epidemics their anticipated punishment, sciolism is the curse of modern humans.

In the course of this study, particular key problems emerged relating to the nature and uniqueness of Byzantine sources. The first problem concerns the reliability of sources, because of three factors: a) the absence of descriptions of plague in Byzantine medical texts, b) the lack of medical knowledge of writers who were no physicians themselves, and, c) the time of writing of the plague descriptions. Regarding the first factor, all kinds of information provided by the sources are the result of writings by the chroniclers and historians of the time and not by physicians. This phenomenon is particularly interesting, since the surviving texts of eminent Byzantine physicians from the sixth to fifteenth century contain no clinical picture of plague. As for the second factor, except for the few cases of writers who—in the broader context of contemporary—possessed basic medical knowledge, the vast majority of them captured a simplistic image of an epidemic on the margins of eschatology and divine wrath. This eschatological predisposition of contemporary writers highlights another problem: their descriptions are characterized by pessimism, uncertainty, and despair. The perpetuation of this literary standard gradually and inevitably led to an established model for describing the disease, which frequently perpetuated stereotyped phrases stemming from past epidemics and may ultimately have nothing to do with the epidemic under study.

As far as the time of writing is concerned, the sources fall into two subcategories: descriptions by eye-witnesses, and descriptions that were drawn up many years after the event. In the latter case, there is an underlying risk of incomplete description, either because the writer did not recall all the events, or because the writer did not experience past events in person, but relied on second-hand information. At this point, the subjectivity of the chronicler plays a catalytic role. It is particularly important to notice that the terms 'pestilence' and 'plague' were gradually assimilated into the nosological entity of plague, despite the fact that, semantically, they might have been associated with an epidemic by any infectious etiological factor.

Another problem lies in the description of plague symptomatology. In several cases the clinical picture is unclear or incomplete, and problematic points are also identified in the differential diagnosis. Nevertheless, whenever described in detail by the Byzantines, the clinical picture appears compatible with the familiar symptomatology. Byzantine writers convey the medical legacy of the ancient Greeks, but differential diagnosis based on the terms βουβών (vouvōn) 'bubo' and ἄνθραξ (anthrax) 'carbon' has proven rather complex. For example, the word 'bubo' itself predisposes the reader with regard to its etymology, thus leading our thought to bubonic plague. The term ἄνθρακας 'carbon' was not related to the modern concept of the infectious disease anthrax—instead, it was simply used to describe the general dermatological manifestation of various diseases—not necessarily infectious.

The difficulty of an *ex post factum* diagnosis has to be taken for granted. This particular difficulty lies in a set of problems that emerge in all cases of studying diseases of the past based on their clinical picture. Perhaps the *Plague of Athens*, known as the *Thucydides syndrome* in medical terms, is the most striking example. Many historians and physicians tried to unravel and identify the disease with the help of the detailed symptomatology offered by the contemporary Greek historian Thucydides. It is really impressive that, based on the reading and study of the symptomatology of the *Plague of Athens*, more than 25 different possible diseases have been proposed, among them bubonic plague, typhus, typhoid fever, smallpox, ergotism, influenza, Ebola virus disease, and many more. Of course, when referring to those times, our diagnostic capability is definitely determined by, and broadly entails, the factor of probability.

Another point of friction when examining a proposed diagnosis often stems from the argument that no disease of the past can be safely defined as plague, since it is not the same disease from the same microbe that we know of today, considering the rate and number of microbial mutations. Whether the human body will get infected, however, depends not solely on the aggression of a new mutant microbe, but also on the defense level of the body. Microorganisms evolve, but the same is not true for defense mechanisms. This simply implies that the defense mechanisms of humans have been invariant and fixed against the invasion of all microorganisms for millennia. At the molecular and cellular level, scientific data support the view of a slow biological evolution of *Homo sapiens*. We can legitimately assume that the reactions and defense of the *Homo sapiens* against pathogens have essentially remained unchanged in the past millennia. The most obvious proof thereof lies in the fact that the impressive decline in mortality due to infectious diseases did not occur earlier than the twentieth century, when antibiotics were discovered. Although the human genome has undergone numerous mutations over the course of millennia, these mutations have not been able to dramatically alter the way in which the body responds to external bacterial stimuli. Before the discovery of antibiotics, humans of the sixth or the early twentieth century became infected with *Yersinia pestis*—regardless of bacterial mutations—and ultimately died in exactly the same way. Mutated or not, microbes have been causing the same sequence of reactions, through the activation and involvement of the same organs (bone marrow, blood, thymus, lymph nodes, spleen) and the same cellular elements (B- and T-lymphocytes, mononuclear macrophages, Natural Killer Cells, and so on).

As mentioned above, the descriptions offered by chroniclers are dominated by a spirit of pessimism, insecurity, and desperation. Phrases like 'there were not enough people alive to bury the dead' are to be found again and again over time—although in several cases they may depict reality, a potential lack of living people could be explained in simple terms by a massive exodus from a city. On the other hand, we cannot contest the exactness of this phrase when considering the high mortality rates of plague. Nevertheless, the lack of reliable demographic data for Byzantine cities raises questions as to the actual rate of diffusion and mortality of diseases. This essentially leads to the question of change in, and progress of, the nature of infectious diseases in that period. Moving from Antiquity to the early Middle Ages, the question arises as to whether a certain change in the susceptibility of the European population against infectious diseases is to be found. Susceptibility or resistance of a population against a particular infectious disease is termed by the collective immunity (or 'herd immunity' or 'community immunity'). It is this resistance that decisively influences the epidemiological behavior of a particular infectious disease in a given population. After all, it is known that the level of immunity declines over the course of years, although at varying rates, while the composition of the population varies accordingly as a result of the addition of new individuals through population shifts. Furthermore, the determination of the 'critical number' of immune individuals, that is, the number of individuals who define the level of collective immunity that suffices to protect a population from the spread of an infectious agent, is of particular significance.

In the broader context of the aforementioned problems, the current study attempts—among others—to expand the way of approaching the primary sources examined in medico-historical literature. Upon reading the primary sources, we ought to consider both *what* is mentioned in the source and *why* it is mentioned. Apart from medical and historical parameters, we understand the way of thinking of the time, as well as the psychology of the writers, since they were both shaped by a set of contemporary social, political, and religious factors.

We have to remember that Historical Epidemiology is a field where numerous theories can be proposed, which have more (or less) probability of being close to the facts. The underlying purpose of this study is to present to non-specialists the basic medical and epidemiological parameters of the disease in the simplest, clearest, and most comprehensible way. An elementary code of communication will be established between scientific disciplines, allowing non-specialists to understand the way scientists with medical knowledge obtain data to build an interpretative historico-epidemiological theory on past epidemics. This method could help scholars to understand and evaluate the medical data related to infectious diseases contained in historical sources. Moreover, this simplified approach can shed light onto the general behavior of microbes, particularly onto the traditional relationships between the factors involved in the emergence of an epidemic. By learning about the evolution of microbes—in our case of *Yersinia pestis*—historians could better understand the possible relationship between this evolution and the historical periods and facts.

Non-specialists cannot be expected to become familiar with medical fields like Molecular Medicine. The core idea of this study is to make the reader aware of the perpetual process of evolution and replication of the epidemiological relationships between humans, environment, and the microcosm. Conversely, references to the historical context of plague epidemics could help specialists to gain a better understanding of historical sources and descriptions of which they may not be aware, whereas these sources may offer useful information for the medico-historical analysis. Readers will definitely be puzzled by the complexity of the relationship between the factors responsible for triggering an epidemic, considering that this is a structured and established relationship that emerges time and again.

Each scientific field draws different data from available historical sources, according to its areas of interest. Instead of bringing individual fields together, however, this anticipated phenomenon may eventually alienate them. It is clear that the purpose of medical research is different from that of historical study. Indeed, medical research on past epidemics has been carried out from the perspective of Molecular Medicine, a field that can hardly be understood without medical knowledge. In light of this obstacle, molecular concepts are presented within a broader framework that may appear trivial or over-simplified to a physician, but is crucial for other readers to be able to understand the basic elements of evolution, and how they relate to the evaluation and approach to historical sources from a medico-historical perspective.

It is also understood that the findings of molecular research offer Medicine significant knowledge of the development of microorganisms, which is not limited to the academia, but entails practical benefits in the treatment of, and protection against, infectious diseases. A well-dated historical source concerning an epidemic that affected a city, the potential archaeological findings of human skeletal remains from the same period and the same city, and a potential finding among these remains related to a pathogen using molecular methods, do not necessarily translate into *de facto* proof that the microbe mentioned in the molecular findings is responsible for the epidemic. Unfortunately, the more distant the past, the less likely it is to reach definitive conclusions backed by irrefutable evidence. This scenario, however, shows that the study of past epidemics, regardless of its starting point, can serve as a meeting point for distinct theories and scientific fields, bridging their differences and fostering cooperation.

Chapter 1
What is Plague?

The Discovery of *Yersinia pestis*

Plague is an infectious disease of rodents which, through their fleas, transmit the causative pathogen of *Yersinia pestis* (henceforth *Y. pestis*) to other animals and to humans. Plague is contagious and its Basic Reproduction Number is R0 = 1.3 but it can also rise to 3.[1] The Basic Reproduction Number is defined as the expected number of secondary cases that one case generates over the course of the transmission period of a disease in a population of susceptible individuals. If R0 is smaller than one (R0 < 1), then each infected individual will transmit the disease to less than one individual and thus the disease will fade out. Conversely, in the case of pathogens with R0 > 1, there is an increased risk of an epidemic spreading in a population.[2]

The bacterium *Y. pestis* was discovered in 1894 by Alexandre Yersin (1863–1943), a Swiss with French citizenship, and the Japanese Shibasaburo Kitasato (1852–1931), in the midst of the terrible epidemic of Hong Kong during the Third Pandemic of plague.[3] This Pandemic (1855–1959) caused the death of approximately 2.2 million people. Originated in the province of Yunnan in southwest China, it spread beyond the borders of China and reached the southeast coast of the country. In February 1894, plague spread to the province of Guangzhou (Canton), and caused more than 70,000 deaths.[4] In the same year, plague broke out in Hong Kong, and spread all over the world through the sea trade routes.

In May 1894, the British authorities of Hong Kong sent a desperate message to the outside world asking for urgent help in dealing with the epidemic which had spread tremendously. Two young microbiologists traveled to the city. The two major European schools of thought that represented the two most renowned schools of Microbiology, namely the Pasteur Institute and the Berlin Institute of Health, essentially clashed. Yersin was a student and colleague of Louis Pasteur, whereas Shibasaburo Kitasato had been taught the secrets of the microcosm by Robert Koch (Nobel Prize, 1905) and Emil von Behring (Nobel Prize, 1901).[5] In essence, the two young scientists represented two different schools with distinct methodology, technology and know-how, which eventually played an important role in the outcome of the search for the microbe. Both scientists had the merit of recognizing the bacte-

---

1 Kermack 1927; Hethcote 1980; Dietz 1993.
2 Gani 2004; Nishiura 2006: 1059–65; Griffin 2011; Hinckley 2012; Lahodny 2013; Didelot 2017.
3 Rosenberg 1968.
4 Wu 1936: 1–55; Liu 2000: i–iv.
5 *Obituary* 1931.

https://doi.org/10.1515/9783110613636-004

rium at a time when the pleomorphism of *Y. pestis* was still unknown. They distinguished two different forms of the same microorganism. On the one hand, Kitasato and his colleague working together on behalf of the British Authorities in Hong Kong, the Scottish physician James Lowson, discovered a form of *Yersinia* enclosed in a sheath, while, on the other hand, Yersin discovered *Yersinia* in its pure form. Kitasato's culture was contaminated by another microorganism, namely *Streptococcus pneumoniae*, to such an extent that it was difficult to describe the bacterium of plague. Yersin, instead, had a pure culture that allowed him to describe the bacterium more accurately.[6]

Classification of *Yersinia pestis*

Alexander Yersin decided to name the bacterium *Pasteurella pestis* in honor of Louis Pasteur. The name was not accepted by the medical community that supported Kitasato.[7] The new genus to which *Pasteurella pestis* belonged, was named *Pasteurellae*. The gradual discovery of new species of the genus *Pasteurellae* with various biochemical specificities caused confusion as to their classification. In 1944, Johannes Jacobus van Loghem (1914–2005) in Amsterdam, shook the scientific community of the time by proposing the 'secession' of the *Pasteurella pestis* and *Pasteurella pseudotuberculosis* species from the *Pasteurellae* genus, and their inclusion into a new genus, which he named *Yersiniae* in honor of Alexander Yersin. At the same time, he suggested that *Pasteurella pestis* and *Pasteurella pseudotuberculosis* be renamed *Yersinia pestis* and *Yersinia pseudotuberculosis*, respectively. This proposal was eventually rejected by the international scientific community.[8] In 1964, Wilhelm Frederiksen in Copenhagen dared to challenge the contemporary classification of *Pasteurellae* once again by classifying the bacterium *Pasteurella X* (*Bacterium enterocoliticum*), which had been isolated in 1939 by J. Schleifstein and M. Coleman working in New York, and unofficially classified under the *Yersiniae* genus.[9] In 1971, after intensive studies and the discovery of new data, the bacteria *Pasteurella pestis*, *Pasteurella pseudotuberculosis* and *Pasteurella enterocolitica* were renamed *Yersinia pestis*, *Yersinia pseudotuberculosis* and *Yersinia enterocolitica*, respectively, and classified under the new genus *Yersiniae*.[10] Since 1985, the official, definitive and commonly accepted name of the microbe has been *Yersinia pestis*.[11] According to the International Code

---

6 Berdiner 1989; Solomon 1997.
7 Perry 1997.
8 Van Loghem 1944.
9 Schleifstein and Coleman 1939; Frederiksen 1964.
10 Volk 1996.
11 Volk 1996.

of Nomenclature of Bacteria (ICNB), the genus *Yersiniae* was officially recognized and came under the *Enterobacteriacaea* family.[12]

Kingdom:	Eubacteria
Phylum:	Proteobacteria
Class:	Gammaproteobacteria
Order:	Enterobacteriales (Enterobacterales)*
Family:	Enterobacteriacaea (Yersiniacaea)*
Genus:	Yersinia*
Species:	***Yersinia pestis***
Subspecies/Biotypes/Strains:	
Y. pestis Angola	
Y. pestis Antiqua	
Y. pestis biovar Antiqua str. B42003004	
Y. pestis biovar Antiqua str. E1979001	
Y. pestis biovar Antiqua str. UG05−0454	
Y. pestis CO92	
Y. pestis FV-1	
Y. pestis KIM	
Y. pestis Mediaevalis str. K1973002	
Y. pestis Microtus str. 91001	
Y. pestis Nepal516	
Y. pestis Orientalis str. F1991016	
Y. pestis Orientalis str. IP275	
Y. pestis Orientalis str. MG05−1020	
Y. pestis subsp. altaica	
Y. pestis biovar Medievalis subsp. Altaica	
Y. pestis subsp. caucasica	
Y. pestis biovar Antiqua subsp. Caucasica	
Y. pestis subsp. hissarica	
Y. pestis biovar Medievalis subsp. Hissarica	
Y. pestis subsp. pestis	
Y. pestis subsp. ulegeica	
Y. pestis biovar Medievalis subsp. Ulegeica	
Y. pestis subsp. talassica	

*Different classifications were proposed according to the new data based on modern genome phylogeny and molecular characteristics. A proposal was made for the order '*Enterobacteriales*' and the family "*Enterobacteriacaea*". The order should be named "*Enterobacterales* ord. nov." which consists of seven families, with, among them, "*Yersiniaceae* fam. nov.". Also, the representatives of the genus *Yersinia* are part of a distinct group of varia genera and their group is the new *Yersinia-Serratia* clade.[13]

---

12 Skerman 1980; Williams 1984; *Judicial Opinion 60* 1985; Wayne 1986.
13 Williams 2000; Parkhill 2001; Deng 2002; Qi 2016; Adeolu 2016.

Geographical Distribution of Plague

Plague has traditionally been under the microscope of the World Health Organization. *Y. pestis* can be found in all continents excluding Oceania. The known endemic foci of the disease are located in the tropical and subtropical zones of the globe, with latitude that is 55° North and 40° South of the Earth's equatorial plane.[14] These latitude regions include many areas where no such endemic foci exist, especially deserts and mountainous areas where few or no rodents survive. The fast detection and identification of the disease enables the timely onset of treatment, thus drastically reducing mortality. The disease has not spread and the risk of a pandemic is currently low. Instead, individual cases occur that are limited to the already known endemic areas of the planet.

In every continent, these endemic areas are associated with a specific type of natural environment and climate, and located in various types of ecosystems (tropical forests, pampas, savannas, steppes). The natural foci are dynamic systems influenced by factors like climate, soil, geography and the fluctuations in the populations of rodents and fleas. Based on the current epidemiological data, plague foci are found in the following areas:[15]

- Europe: natural plague foci are located on the outskirts of the Caspian Sea basin and the eastern parts of the Caucasus.
- Euro-Asiatic zone: the boundaries of this disease focus are located north-westwards of the Russian steppe, that is, between the rivers Volga, Don, and smaller rivers of the Ural Mountains, and south-east of the Gobi desert.
- Asia: the endemic regions of plague are located along the southern border of the steppe, and extend from the foothills to eastern Turkey to northern Iran and up to northeast China. Other endemic foci are to be found in Cambodia, India, Indonesia, Iran, Mongolia, Nepal, Myanmar (formerly Burma) and Vietnam.
- Middle East: the areas are located in the south of the Arabian Peninsula, and more specifically on the Saudi Arabia-Yemen border.
- Africa: several zones in the African continent have been identified as endemic, which are located deep in the tropical forests, and may belong to neighboring countries (such as the lakes Albert and Edward at the Uganda-Democratic Republic of Congo border). Endemic foci can be found in various countries: the Democratic Republic of Congo, Kenya, Libya, Madagascar, Mauritania, Mozambique, Namibia, Senegal, South Africa, Tanzania, Uganda and perhaps Egypt.
- America: in North America, natural foci can be found in 15 Western states, as well as on the USA-Canada and USA-Mexico borders. In South America, foci have been reported in Argentina, Bolivia, Brazil, Ecuador, Peru and Venezuela.

---

14 Poland 1999.
15 Poland 1999.

In the late nineteenth century, humanity was once again afflicted by plague. During the Third Pandemic, the disease spread from Guangzhou (Canton) and Hong Kong to the whole world through the maritime trade routes. During the period 1894–1903, the steamships that had replaced the old, slow-moving sailboats, introduced the disease to 77 ports throughout the world: 31 in Asia, 12 in Europe, 8 in Africa, 4 in North America, 15 in South America, and 7 in Australia. At the dawn of the twentieth century, plague had spread around the globe. The disease appeared in countries which had never been affected before, while the old and forgotten foci were reactivated. A first general picture of the prevailing situation with regard to the disease was offered by Knud Stowman (1889–1978) in 1945.[16] In 1954, the WHO published the first results of multiannual studies on epidemics in 46 countries covering the period from 1900 to 1950.

According to the data of the WHO, a total of 38 countries reported outbreaks or isolated cases in the period 1957–2003. Overall, 94,248 cases and 8,740 deaths were reported worldwide. The gradual reduction of plague outbreaks is due to the improvement of living conditions and health services in some countries, as well as to the drastic decline of the dangerous pneumonic form of plague. Based on the data for the period 1957–2003, it appears that 58.4% of reported cases occurred in Asia, 27.8% in Africa and 13.8% in the Americas.[17] The last pneumonic plague epidemics were recorded in 2006 in Congo, where three epidemics that lasted from June to November resulted in 1,900 patients and 111 deaths.[18] In 2017, Madagascar endured an outbreak of plague with a total of 2,147 confirmed cases and 209 deaths.[19]

According to the WHO doctrine, the control of hosts has to be the absolute priority.[20] This was the basic principle proposed in 1947 by John E. Gordon and Philip Thomas Knies, which the WHO embraced. Based on this principle, the health authorities must follow an aggressive tactic and focus their efforts on exterminating the hosts of natural foci (focal attack). The control of dead rodents that carry a large number of ectoparasites (fleas) is a part of this framework. This constitutes a very significant research topic, as it is believed that the presence of this type of flea helps to spread the epidemic more quickly. Therefore, the management of dead rodents in declared natural foci is of outmost importance.

Morphology of *Yersinia pestis*

Y. pestis is an intracellular, Gram-negative cellular bacterium, with an average size of 1–3 μm in length and 0,5–0,8 μm in width. It is non-motile and uses aerobic respi-

---

16 Pollitzer 1954: 71–72.
17 Poland 1999.
18 Bertherat 2005.
19 Nguyen 2018.
20 Gordon 1947; Gratz 1983.

ration to reproduce, but it is characterized as a facultative anaerobic bacterium. As already mentioned, ever since the bacterium was discovered by Yersin and Kitasato, its morphology was problematic, since the two researchers observed two distinct forms of the same microorganism. Its typical morphological picture is that of a small spherical bacillus. Its size and shape vary widely and can take the form of a small grain or stick. It is also found in a thin or thick filamentous form.[21] It has been observed that this pleomorphism is more prevalent when the amount of salt in the nutrient medium increases.[22] In material gathered from swollen lymph nodes, various bacterial cell formations are stained blue, while other formations appear in light cyan or red, and the whole cell has a safety pin shape.[23]

The first studies referred to a sheath surrounding the bacterium. In 1914, however, Rowland came to the conclusion that the sheath was not a fixed morphological feature of the bacterium. Researchers eventually concluded that the sheath did not appear under all conditions—although it was a typical characteristic of the bacterium—and they thus realized the great pleomorphism of *Yersinia*.[24] The forms entailed in the sheath mainly appear at temperatures of 33°C–38°C.[25] The studies of the bacteriologists Heinrich Albrecht (1866–1922) and Anton Ghon (1866–1936) revealed that filamentous forms appear in nutrient environments containing glycerol or sugars. At the Institut Pasteur in Madagascar, Gustave Bouffard (1872–1957) and Georges Girard (1888–1985) also demonstrated that low temperature incubation conditions lead to the formulation of longer forms.[26]

Growth Conditions, Survival and Resistance

Y. pestis has the ability to grow in a wide range of temperatures between 4°C and 40°C, but especially at 28°C–30°C and at pH ranging from 5.0 to 9.6 and especially in neutral-slightly alkaline environments (7.2–7.6).[27] Under specific circumstances, temperature can range from -2°C to +45°C.[28] The bacterium grows when the oxygen in the nutrient medium ranges from 0.72 to 0.75 ml/100 ml of the nutrient substrate. When the oxygen content reaches 0.04 ml/100 ml, which usually takes place within 21 and 24 hours, the bacterium ceases to grow, while after 27 hours the logarithmic (or exponential) phase of growth of *Y. pestis* begins.[29]

---

21 Palmer 1994.
22 Williams 2000; Parkhill 2001; Deng 2002; Qi 2016.
23 Joklik 1992; Cornelis 2002.
24 Pollitzer 1954: 71–72.
25 Brubaker 1972; Barnes 1992; Poland 1994; Brooks 1998: 253–255.
26 Bouffard 1923.
27 Dennis 2000.
28 Dennis 2000.
29 Staggs and Perry 1991.

When agar SBA (Sheep Blood Agar) is used, which is a medium employed for isolating and examining the morphological characteristics of a colony, cultures can grow at two different temperatures, namely at 28°C and 37°C.[30] The growth of Y. pestis is ideal and rapid at 28°C, but the critical temperature is that of 37°C, since the F1 thermoregulatory antigen expresses at this temperature, which is an antigen related to the mechanism of the disease and constitutes a characteristic feature for diagnosing the bacterium.[31] Moreover, the very first studies revealed the importance of amino acids, sulfur compounds (S^{2-}, $S_2O_3^{2-}$, SO_3^{2-}) and iron for the development of the microorganism.[32]

The bacterium is destroyed by heat at 56°C in 15 minutes, or by exposure to sunlight for four hours.[33] Since the discovery of the bacterium, the effect of the free environment on the transmission of the disease has been the subject of numerous studies. It gradually became apparent that the free environment does not play a particularly important role in the transmission of the pneumonic or bubonic form of the disease. In 1934, Wright discovered that Y. pestis was rapidly eliminated when it entered in direct contact with blood serum or gelatin in aqueous solutions.[34] In 1944, Girard announced that the bacterium could survive at temperatures lower than 25°C without losing its virulence for a whole week. At temperatures between 2°C and 4°C, Y. pestis was able to survive for more than two years.[35] Recent studies suggest that, on the surface of objects and under specific circumstances, Y. pestis can survive and remain infective for a minimum of five days.[36]

Y. pestis is a bacterium that is easily killed by sunlight. One hour of exposure to sunlight suffices.[37] The duration of exposure to sunlight also depends on the material on which the bacterium is located. Exposure time to solar radiation may need to extend for up to four hours. The bacilli in sputum of patients with pneumonic plague are killed in two to five hours, but this can take up to 12–14 hours, depending on the circumstances.[38]

The dry environment is not considered capable of killing the bacterium, as it typically takes a temperature of more than 100°C for one hour to definitively kill Y. pestis. The sterilization of various objects is achieved by heating at 55° C–60° C for 15 minutes or at 65°C for one hour.[39] Research has shown that the bacterium is not susceptible to low temperatures, but, on the contrary, it is surprisingly resilient. Even in

---

30 May 2001.
31 Smith 1996; May 2001.
32 Pollitzer 1954: 80; Poland 1994.
33 Palmer 1994; Smith 1996.
34 Pollitzer 1954: 78–79.
35 Rose 2003.
36 Rose 2003.
37 Poland 1994: 104.
38 Palmer 1994.
39 Palmer 1994.

cultures preserved at temperatures of ± 2°C or ± 4°C, the growth of the bacterium continues, though at a slower pace. Cultures preserved for a period of five and a half months at -31°C showed decreased infectivity but remained active.[40] In general, low temperatures can be said to have a positive effect on the preservation of *Y. pestis*.

An issue of particular importance when studying *Y. pestis* is the impact of symbiotic microorganisms. The techniques used to make a culture and the methods for identifying the bacterium reveal the existence of a clear relationship between *Y. pestis* and other microorganisms. This relationship can be either symbiotic or antagonistic.[41] In the 1930s it was established that the coexistence of *Y. pestis* and *Mycobacterium tuberculosis*, the causative agent of tuberculosis, does not affect their parallel development *in vitro*.[42] As a researcher of the Institute Pasteur in Madagascar, Girard realized that during the plague epidemics that affected the island between 1922 and 1936, the patients of the local leper house had never been infected with plague. Experimenting with guinea pigs that he had infected with the bacterium *Mycobacterium lepraemurium*, the causative agent of leprosy in rodents, he observed that none of them became infected with plague and thus he formulated the concept of cross-immunity.[43] Studies so far have indicated numerous *Y. pestis* antagonists, such as the bacteria *Alcaligenes faecalis* (*Bacillus faecalis alcaligenes*), *Bacillus mesentericus*, *Chromobacterium prodigiosum*, *Proteus vulgaris*, *Pseudomonas pyocyanea*, *Streptococcus mucosus* and *Vibrio cholerae*. Conversely, *Bacillus aromaticus*, *Bacillus subtilis*, *Klebsiella ozoenae*, *Staphylococcus aureus* and *Vibrio paracholerae* are considered *Y. pestis* symbiotic bacteria.[44] It is worth mentioning that various bacteria that cause dysentery occasionally appear as antagonistic or symbiotic to *Y. pestis*.

Summarizing previous and more recent scientific data, we may conclude that the bacterium of plague can—under specific conditions—survive outside living organisms. However, its survival time in the environment, as well as its virulence, depend upon numerous factors. Under specific conditions, *Y. pestis* can also survive in inanimate objects, but whether it is infective remains to be established.

Infectious Agents of *Yersinia pestis*

During an infection process, a bacterium is exposed to a series of stimuli that often make its survival improbable. The bacterium of *Y. pestis* reacts to the defense mechanisms of the human cells with the help of its infectious agents. These agents invade

---

40 Poland 1994: 105.
41 Poland 1994: 108–109.
42 Pollitzer 1954: 108–109.
43 Girard 1951; Girard 1975; Ell 1987.
44 Pollitzer 1954: 108–109.

the cells of the host, and proliferate and increase the resistance of the bacterium to phagocytosis, that is, to the counterattack of the host cells.[45]

The pathogenic activity of *Y. pestis* is conditioned by the presence of infectious agents like Ca^{2+} level, antigens O, V and W, the *Yersinia* outer proteins (Yops), the Fraction 1 (F1) antigen, the plague toxin, as well as other specialized and complex mechanisms.[46]

At least 20 different antigens have been detected and analyzed thus far. It has now been proven that many of them are common to *Y. pseudotuberculosis* and *Y. enterocolitica*, a fact that probably is of particular significance in the history of the disease.[47] One of the major weapons that the microbe possesses is the F1 surface antigen, which is located on the cell surface of the bacterium. The F1 is a temperature-dependent antigen that appears when the bacterium develops at temperatures $\geq$ 33 °C, but primarily at 37 °C.[48] The Yops (*Yersinia outer proteins*) system is another highly specialized mechanism of the microbe that illustrates the high level of evolution of the microorganism. The Yops system plays a catalytic role in the resistance of *Y. pestis* against the phagocytosis from the cellular components of the immune system, namely the macrophages that engulf the invader.[49] Although this mechanism has not been fully explained, it is considered certain that the F1 antigen and the Yops protein system act together to make *Y. pestis* resistant to phagocytosis.[50] More data will be presented in the following chapters.

Epidemiological Cycle of *Yersinia pestis*

On June 2, 1898, the envoy of the Pasteur Institute in Karachi, Paul-Louis Simond (1858–1947), carried out a historical, yet extremely dangerous experiment. He placed two cages containing two mice next to each other: the first mouse was already infected with plague, while the second was healthy. In the cage containing the infected mouse he also placed fleas which were able to move to the cage containing the healthy mouse. A few days later, the healthy mouse presented the symptoms of plague. The discovery that an insect could carry a bacterium was incompatible with the data and perceptions of that time when—despite the impressive discoveries in the field of microbiology—medicine was still embedded in the theory of 'miasma' and 'bad air' that spread diseases among humans.

The epidemiological cycle of plague originates in wild rodents that constitute the natural endemic reservoir of the disease. Plague is transmitted to humans through

---

45 Volk 1996.
46 Joklik 1992.
47 Perry 1997.
48 Zhou 2006.
49 Cornelis 2002.
50 Cornelis 2002; Du 2002; Kerchen 2004.

two epidemiological cycles which are often linked to each other. In the sylvatic plague the microbe is transmitted among wild rodents or other forest species like squirrels, hares, or wild rabbits. The transmission between animals can occur via fleas, or the presence of fleas on the ground (for instance, when a species occupies the nest of another species). Moreover, the transmission of sylvatic plague can occur when a species devours another that is its natural enemy. At this point, it should be mentioned that humans enter this epidemiological cycle mainly because of activities like hunting, gamekeeping, or leather processing, for example.[51]

In rural, semi-urban or urban areas, the fleas infected with *Y. pestis* can transmit sylvatic plague to rodents which, in turn, transmit the disease directly to humans (via excoriations) or via their fleas (urban epidemiological cycle). The 'urban' rodents more often infected are *Rattus rattus* and *Rattus norvegicus*, the geographical distribution of which is global. Urban epidemics are historically related mainly to the epizootic of *R. norvegicus* and *R. rattus*.[52] A significant epidemiological role is also played by the rodents *Tatera brantsi* and *Mastomys natalensis* (which are also responsible for causing Lassa hemorrhagic fever) that are involved in the plague epidemics on the African continent.[53]

The spread of the disease is also influenced by the eating habits of some people. It has been proven that *Y. pestis* can be transmitted to humans by eating the meat of camels or desert rodents (marmots). In these cases, the disease occurs with pharyngeal or gastrointestinal symptoms.[54] The transmission of the disease between humans is possible mainly through the droplets of pneumonic plague patients or through the inhalation of microbial dust (suspension), while the latter mode of transmission is mainly responsible for the primary pneumonic form of the disease in humans.

Approximately 200 different rodent species are involved in the spread of sylvatic plague. They are to be found in areas of the former Soviet Union, the Southwestern United States of America, India, Vietnam and Africa.[55] Epizootic is greatly influenced by mice, rats, gerbils, squirrels, marmots, raccoons and coyotes.[56] Moreover, it has been determined that the disease can be transmitted to pigs too, as witnessed in 1899 and 1900 by the Austrian-German Commission on Combating Plague.[57]

The disease affects rodents mostly in the spring and summer, while the incidence of enzooty fluctuates on an annual basis. Epizootic is cyclically regulated by the number of the animals that eventually die, by the resistance to the disease of the animals that survive, and by the fleas' search for new, less resistant hosts.[58] As far as

---

51 Poland 1973.
52 Dennis 2000.
53 Green 1978.
54 Christie 1980.
55 Palmer 1994.
56 Velimirovich 1990; Poland 1994.
57 Albrecht 1900.
58 Fordham 1976.

felines are concerned, the disease also affects domestic cats, the mortality of which stands at 56%.[59] With regard to canines, the disease infects foxes, wild dogs and coyotes, and cases of epizootic have also been reported among bears (*Ursidae*).[60] A highly interesting topic currently under study and monitoring is the transmission of the disease to animals during hibernation. It has been observed that, when an animal becomes infected during hibernation, it does not exhibit the symptoms of plague before the end of its hibernation.[61]

As already mentioned, the flea is a vector in the epidemiological cycle of plague. At this point, it is necessary to briefly examine this insect. Fleas are involved in approximately 80% of human-related cases (the rest 20% are of cases of infection through contact with rodents due to excoriations).

Fleas belong to the order of *Siphonaptera* and are part of the *Insecta* class. The order of *Siphonaptera* includes approximately 2,500 species, of which 200 are of medical and veterinary interest. Their most common genera belong to the *Pulicoidea* and *Ceratophylloidea* superfamilies. The *Pulicoidea* superfamily is very interesting from an epidemiological perspective as it includes the *Xenophyllinae* fleas, that is, the family to which also the *Xenopsylla cheopis* belongs. High temperatures and low humidity have a significant effect on the life cycle of fleas, and each species has its own environmental and climatological preferences. For instance, *Xenopsylla cheopis* prefers warm and humid climates, *Xenopsylla brasiliensis* humid ones, and *Nosophyllus fasciatus* dry climates. Temperature plays an important role in the hatching of flea eggs, which can withstand a wide range of temperatures, from $-13C^0$ to $+34C^0$ for *Xenopsylla cheopis*, from $-8C^0$ to $+34C^0$ for *Pulex irritans*, and from $-5C^0$ to $29C^0$ for *Nosophyllus fasciatus*.[62] The female flea lays its eggs mainly in the nests of its hosts, but also in the environment. The eggs are small, smooth, and covered by a thin sticky coating. The number of eggs varies from species to species—for example, *Xenopsylla cheopis* lays 300–400 eggs, *Pulex irritans* 440–450 and *Ctenocephalides felis* 800 eggs.[63]

The development of larvae takes 10 to 21 days, but, under unfavorable conditions as the lack of food or low temperatures, it can be extended up to 200 days.[64] At the pupal stage, development continues within a cocoon that is formed by a substance secreted by the salivary glands of the mature larva. The emergence of the adult flea from the cocoon occurs by a stimulus, such as vibrations during the movement of hosts, or the increase in the humidity of the environment. The lifespan of an adult is usually between ten days and six weeks, a period that may extend up to one year under specific circumstances. The size of adult fleas ranges between 1 and

---

59 Eidson 1988.
60 Rust 1971; Marshall 1972a; Mann 1979.
61 Palmer 1994.
62 Twigg 1978.
63 Pollitzer 1954: 315–316.
64 Mears 2002.

4 mm. Fleas have a flat-bodied appearance, usually with a dark brown colour. They are wingless, with three pairs of well-developed legs. The third pair of legs is more prolonged and allows for large jumps which are disproportionate to the size of the insect (kangaroo movement), and can reach 20 and 30 cm (or even more) in height and length, respectively.[65]

When a flea bites an infected rodent, it takes in a quantity of microbe along with the blood. Within three to nine days after taking a blood meal from an infected animal, the bacterium grows and eventually blocks the proventriculus of the flea. The proventriculus is a valve-like anatomical structure located between the midgut and esophagus.[66] Once this structure is blocked, the flea becomes starved for blood and more aggressive, thus trying to feed more often.[67] The flea sucks blood with each new bite, but, due to the blocked proventriculus, the blood flows back to the wound from the bite, and the flea appears to be vomiting. As a result, the blood that flows back to the wound carries a quantity of microbes from the proventriculus of the flea. A flea bite can transmit up to 20,000 bacteria to the host.[68] The bacteria are then transferred from the wound to the regional lymph nodes. When the rodent dies, the fleas abandon its body in search for alternative hosts among humans and other mammals, or remain in the free environment or in the ground for some time. Under ideal temperature and humidity conditions, the fleas can survive and remain infected for at least six weeks, although they may be incapable of transmitting the disease after 14 days.[69]

Finally, in addition to the established role of fleas in the epidemiological cycle, some other insects are implicated as potential vectors: lice (*Pediculus humanus capitis*), bed bugs (*Cimex lectularius*), mites that belong to the *Tyroglyphidae* family and feed on infected rodents (*Glyciphagus domesticus* and *Tyroglyphus siro*), ticks (*Rhipicephalus haemaphysaloides*, *Argas persicus*, *Dermocentor silvarum*, *Hyaloma volgense*, *Rhipicephalus sanguineus*), mosquitoes (*Mansonia*, *Culex pipiens*, *Anopheles rossi*, *Aëdes aegypti*), beetles (*Sitophylus*, *Tenebrio molitor*) and cockroaches (*Periplaneta Americana*, *Rhyparobia maderae*, *Blatella germanica*). The role of ants is still under investigation.[70]

Differential Diagnosis, Clinical Picture and Therapy

In the first days after the onset of the disease, differential diagnosis is difficult as various infectious diseases have clinical symptoms and signs that are similar to those of

---

65 Rothschild 1972.
66 Cavanaugh 1971.
67 Burrougs 1947.
68 Burrougs 1947.
69 Twigg 1978.
70 Macchiavello 1949; World Health Organization 1950; Strand 1977.

plague. Upon the appearance of swollen lymph nodes, differential diagnosis must be used to distinguish from other lymphadenopathies as tularemia, lymphogranuloma venereum (LGV) and/or common lymphadenitis caused by streptococci and staphylococci. Other common diseases that are often involved in differential diagnosis are infectious mononucleosis, filariasis, typhus, anthrax, leptospirosis, legionnaires' disease, and Debre-Foshay-Mollaret-Reilly disease (cat-scratch disease). During the phase of suppuration of the lymph nodes, however, the range of other etiological factors is significantly limited. Sometimes the effect of swollen lymph nodes in the mediastinum and abdomen is evident. In these cases, the pain in the abdominal region should be distinguished from the pain caused by appendicitis, colitis and cholecystitis. In the cases of primary pneumonic plague, differential diagnosis entails almost all forms of the lower respiratory tract. In patients with pharyngeal plague, differential diagnosis is performed in relation to streptococcal or viral pharyngitis.

The incubation period of the disease lasts from one to twelve days, but usually this time spans two to four days. More specifically, the incubation period of bubonic plague is two to eight days, whereas, for primary pneumonic plague, the incubation period ranges between one and six days. After the incubation, plague appears in an acute and often fulminant manner and may take various forms: glandular (bubonic), pneumonic, septicemic, pharyngeal or plague meningitis. A thorough examination of the skin reveals the point of the bite by the flea, which usually goes unnoticed by patients.

Glandular (bubonic) plague occurs with sudden symptoms such as chills, high fever (38.5°C–40°C), headache, tachycardia, vomiting and severe malaise. Almost simultaneously, several lymph nodes begin to swell. When palpated, the nodes give the impression of a mass under the skin; each of them can range from two to five centimeters, but rarely up to ten centimeters in diameter. Plague affects primarily the inguinal lymph nodes (in 60–80% of the cases), and secondarily the axillary (30%) and thoracic (10%) lymph nodes.

After one to two weeks, the lymph nodes are automatically opened towards the overlying skin, and pus begins to flow out. At this stage, there is a strong hemorrhagic diathesis provoked by the action of the plague toxin on the vessels. This hemorrhagic diathesis causes petechiae and ecchymoses, and hemorrhages of the mucous membranes in the digestive, respiratory and urinary tracts occur in the most severe cases. The face of patients appears pained and agonizing, and their look is wild (*facies pestica*). As the disease progresses, severe symptoms in the nervous system appear, such as slurred speech, delirium, and/or manic excitement. The final stages are accompanied by symptoms of organ failure (multiple organ failure-MOF), adult respiratory distress syndrome (ARDS), and disseminated intravascular coagulopathy (DIC) syndrome, which is evident in 85% of patients.[71]

---

71 Wenzel 1996; Poland 1999: 43–48.

Pneumonic plague can take two forms: primary or secondary.[72] Secondary plague affects 5% of bubonic plague patients. These patients are considered as the main cause of primary pneumonic plague by transmitting the disease through their droplets. Pneumonic plague has an incubation period that ranges between one and six days; it is a severe infection of the respiratory system manifested by high fever, severe malaise, intense coughing, dyspnea, and cyanosis.[73] In secondary plague, these symptoms are associated with the symptoms of glandular plague. The sputum discharged is hemorrhagic and contains the pathogenic bacillus. The prognosis is unfavorable; if left untreated, death occurs within one to five days.

The septicemic form is the evolution of glandular or pneumonic plague. It occurs in 40% of patients under treatment, and in 100% of untreated. It is characterized by intense chills, further rise of the body temperature, increased headaches, vomiting, and delirium.[74] Plague meningitis occurs in 6% of untreated patients and manifests itself with neck stiffness, headache, confusion, and coma.[75] Moreover, pharyngeal plague is a rare form, the symptomatology of which resembles that of acute tonsillitis. It is evident in individuals who have been in contact with patients suffering from pneumonic plague.[76]

A critical point in the course of the disease lies in the early onset of therapy. Modern antibiotics secure the complete recovery of patients, whereas early treatment allows patients to be afebrile within three days after the onset of treatment.[77] The Food and Drug Administration (FDA) of the United States of America has designated streptomycin as the drug of choice for treatment.[78] Other such drugs are gentamicin, chloramphenicol, and tetracycline.

It should be noted that laboratory studies of strains and cultures of *Y. pestis* obtained from patients are carried out according to the Biosafety Level 2 (BSL2) protocol. Particular care has to be taken when performing a necropsy on infected rodents acquired either from the laboratory or from the environment.[79] The Biosafety Level 3 (BSL3) guidelines are required for numerous cultures, operations involving microbial suspensions, or antibiotic-resistant strains.[80] *Y. pestis* belongs to the highest risk biological weapons (Category A), along with *Bacilus anthracis*, *Brucella suis* and *Brucella melitensis*, *Francisella tularensis* and the smallpox (variola) virus.[81] The WHO has developed specific therapeutic regimens to be applied if *Y. pestis* is used as a weapon of

72 Campbell 1998.
73 Inglesby 2000: 2281–2290.
74 Campbell 1998.
75 Wenzel 1996.
76 Marshall 1972b.
77 Campbell 1998.
78 Campbell 1998.
79 Pike 1976; US Department of Health and Human Services 2007.
80 Centers for Disease Control and Prevention 1982.
81 Eitzen 1997.

mass destruction, by dividing patients in three categories: adults, children, and pregnant women.[82]

Dying from *Yersinia pestis*

Upon entering the organism, the Yops proteins of *Y. pestis* disrupt the membrane cohesion of the host cells and interrupt their response, namely phagocytosis; at the same time, they inactivate the function of macrophages. Various bacterial pathogens employ a conserved type of secretion system to inject virulence effector proteins directly into the cells, so as to undermine the functions of the host. *Y. pestis* uses the type III secretion system (T3SS), which is a central element of the virulence of the microbe.[83]

Prior to contact with the target-cell, *Y. pestis* alters its cell surface by creating syringe-like projections (injectisomes) that enable the influx of Yops proteins into the host cell.[84] Upon their invasion of the cell, the Yops proteins act in smaller groups, and each such group takes on a specific role. One group essentially prepares the advent and functioning of the other, until the cytoskeleton eventually becomes disorganized.[85]

As in the case of rodents, the human erythrocytes affected by *Y. pestis* (*in vivo* or *in vitro*) must soon be used as a natural source of iron (Fe^{2+}) and porphyrin intake for the bacterium to survive, proliferate, and grow.[86] In mammals, iron is linked to specific proteins, iron-binding proteins, and hemoproteins, which are most commonly used by *Y. pestis* through a special iron acquisition system.[87] The procedures used in obtaining iron and porphyrin eventually lead to the oxidation of hemoglobin to heme.[88] In addition, some proteins of the microbe disturb the normal functioning of erythrocytes, which become incapable of transferring oxygen.[89]

Decreased transfer of oxygen to the tissues is essentially responsible for a series of clinical symptoms of plague. These clinical symptoms can be compared to those of specific cases of poisoning by toxic substances.[90] The inability of the cell to transfer oxygen results in typical clinical symptoms like headache, vomiting, chills, dizziness, somnolence or coma, and, eventually, cyanosis.[91] These symptoms usually resemble

---

82 Committee of Infectious Diseases 1997; American Hospital Formulary Service 2000.
83 Gophna 2003; Coburn 2007; Plano 2013; Nans 2015.
84 Cornelis 2002.
85 Cornelis 2002; Zhou 2006.
86 Ratledge 2000; Feodorova 2002.
87 Beutler 1978: 603–623.
88 Burrows 1964.
89 Perry 1999.
90 Feodorova 2002.
91 Beutler 1978: 603–623.

those of poisoning by aniline, carbon monoxide or nitrobenzene, and in fact the final result is a progressive tissue hypoxia and cyanosis.[92] The spectrum of the host response may include a pathological event like DIC syndrome and can lead to arteriolar thrombosis, haemorrhage in skin, serosal surfaces, organ parenchyma, acral cyanosis, ischemia and gangrene of acral parts.[93] Gangrene, which is evident by this time, give the body a brownish or black complexion, which is why the disease has been traditionally described as the "Black Death".

---

92 Burrows 1964; Perry 1997.
93 Poland 1999: 43–46.

Chapter 2
Inside the Molecular World of *Yersinia pestis*

Principles of Molecular Biology

Every living organism is similar to its ancestors in most aspects. The preservation of the special properties or the stability of traits is defined as heredity. Prior to the division of cells for the creation of two new cells, the genetic material is organized into a structure called chromosome. In the case of both prokaryotic (bacterial) cells and eukaryotic cells (cells of higher animals and plants), each divided chromosome forms two new identical copies of itself. Nevertheless, during the division process, specific differences exist between the prokaryotic and eukaryotic cells. The processes of mitosis and meiosis exist in eukaryotic cells, but are absent in the division of prokaryotic cells. Although replication or binary fission of prokaryotic cells may appear to be a simpler mechanism compared to the proliferation of eukaryotic cells, it is in itself a complete and accurate hereditary mechanism.

The structural and functional characteristics of organisms are the result of the expression of a series of information inherited by each organism from its ancestors and passed on to its descendants. This information is defined as inherited or genetic, and it is recorded in the sequence of the nitrogenous bases of DNA, namely the molecule that makes up the genetic material of cells. In 1953, James Watson (b. 1928), Francis Crick (1916–2004), and Maurice Wilkins (1916–2004) (Nobel Prize 1962), and Rosalind Franklin (1920–1958), proposed the well-known double-stranded DNA theory, which defines its layout. The course and process by which genetic information stored in the DNA molecules 'flows' towards proteins—which are responsible for the basic structural and functional characteristics of cells—constitute the central dogma of molecular biology. A prerequisite for the transfer of genetic information from parent to daughter cells is the self-replication of the DNA molecule.

Two very similar biomolecules of the cell are responsible for the storage of genetic information, namely the DNA (Deoxyribonucleic acid) and the RNA (Ribonucleic acid). The DNA and RNA consist of four different monomers called nucleotides, each of which is composed of three subunit molecules (a five-carbon sugar molecule, a nitrogenous base, and one phosphate group). A nucleotide of DNA contains one of the following nitrogenous bases: adenine (A), guanine (G), cytosine (C), thymine (T). A corresponding nucleotide in RNA contains one of the following bases: adenine (A), guanine (G), cytosine (C), uracil (U). The nucleotides are linked together in a chain that is also the sequence of DNA and RNA.

https://doi.org/10.1515/9783110613636-005

Eukaryotic Cells

As mentioned above, the process of expression of genetic information is the *central dogma of molecular biology*, which evolves in three stages: i) replication, ii) transcription, and iii) translation.

Replication

The mechanism of DNA synthesis that allows the transmission of genetic information to the next generation is called replication. This is achieved through the unwinding of the double helix. The double-stranded DNA unwinds and forms a single-stranded DNA; then, with the help of the enzyme polymerase III, the two separate clones replicate, resulting in their duplication. Replication follows the principle of complementary base pairing, where the nucleotides containing the nitrogenous bases adenine (A), thymine (T), guanine (G), and cytosine (C) align opposite the nucleotides that respectively contain the nucleobases thymine (T), adenine (A), cytosine (C) and guanine (G), thus forming base pairs (A-T, T-A, G-C, C-G).

Transcription

In the case of eukaryotic cells, given that the DNA is located at the nucleus, whereas the ribosomes—that is, the cell organelles where proteins are synthesized—are located at the cytoplasm, genetic information has to be somehow transferred from the nucleus to the cytoplasm. The solution to this problem is offered by an intermediate molecule which is produced based on the DNA model and transfers the information from the nucleus to the cytoplasm. The existence of this intermediate molecule, which has been termed 'messenger RNA' (mRNA), constitutes the stage of the process called transcription. Upon completion of this process, a single-stranded mRNA molecule has been synthesized, whose sequence of ribonucleotides was 'dictated' by the sequence of deoxynucleotides of the transcribed portion of the DNA.

Translation

The third stage of the process is translation, and concerns protein synthesis by messenger RNA. The proteins are macromolecules that consist of long chains of amino acids (from 50 to 30,000 amino acids) and ensure basic cellular functions like metabolism, replication, or regulation of gene expression. In this process, which is carried out in the ribosomes, the sequence of the mRNA nucleotides dictates in turn, the production of chains with a specific sequence of amino acids. Experiments carried out in the 1960s specified the amino acid encoded by each nucleotide triplet called codon.

Prokaryotic Cells

Due to the absence of the nuclear membrane in prokaryotes, the stages of transcription and translation may be linked together and they are not physically separated. As we have seen, nucleic acids turn into proteins, and this process requires that each cell has a code (genetic code) responsible for this transformation. A full sequence of a cell's genetic information constitutes the genome; it is the genetic heritage of a species. Genetic information of the bacterial cell is stored in a circular chromosome and in some extrachromosomal genetic structures, the so called plasmids. The chromosome is an organized DNA structure that comprises many genes. The plasmids are small circular double-stranded DNA molecules which are independent from the bacterial chromosome. Plasmids contain their own replication mechanism that allows them to replicate independently of the bacterial chromosome. Nevertheless, despite the fact that they are autonomous, they replicate almost simultaneously with the main chromosome. Interestingly, a plasmid-bearing bacterial cell can come into contact with another, plasmid-free bacterial cell. The bacterial donor cell can transfer plasmid to the bacterial host cell, which in turn, can repeat this process as donor, transferring plasmid to a plasmid-free host cell.

Genetic Variation and Mutations

The set of genetic factors carried by a bacterial cell is called genotype. The set of observable characteristics that stem from the genotype is defined as phenotype. Mutations that define changes in phenotypic characteristics also require a change in the genotype. Based on the type of phenotypic change, the changes of genotype can be described as gain-of-function mutations or loss-of-function mutations.

The main source of genetic variation is DNA change (mutations) that alters its sequence of nucleotides and thus the information that it contains. Some changes in the sequence of nucleotides are minor, but others can incur major rearrangements in the sequences, duplication of genes or parts of the genome, and even deletion of parts of the genome.[94]

At this point, some basic principles of molecular biology are necessary. They shall be briefly explained with the help of a model organism in experimental research, namely the *Escherichia coli* (*E. coli*) bacterium which is a classical model for the study and understanding of the microcosm. There is a symbiotic relationship between the bacterium and humans and other mammals, since the bacterium is maintained in the body via substances that escape digestion and absorption in the stomach and small intestine.

---

94 Alberts 1998: 289 – 317.

E. coli is ideal for research due to the basic advantage of bacteria: it only contains one copy of its genome, which means that it is haploid. On the contrary, human cells contain two copies of their genome (diploid). Therefore, when a bacterial gene undergoes a mutation, a detectable hereditary change in its characteristics may occur, that is, in the bacterial phenotype. On the other hand, in diploid organisms—that is, the type of organisms to which humans belong—a mutation does not necessarily lead to changes in the phenotype, because each cell also contains a non-mutated copy of the same gene which often counterbalances the effects of the mutated gene. Bacteria reproduce by binary fission; their DNA is replicated and separated at the two ends of the bacterium. The bacterium is then divided into two daughter cells, each of which contains the same genome as the parent cell, provided that no mistakes were made during the process of replication.

In vitro, *E. coli* grows rapidly in cultures and doubles every 20 – 25 minutes. The high speed at which bacteria are divided means that mutations occur within a short period of time. Moreover, bacterial mutations can result from the change of environmental conditions. Once again, *E. coli* can help us understand the role of the environment, as it is sensitive to the antibiotic rifampicin. Among the high numbers of bacteria in a culture (10^9 cells), however, it is possible that some antibiotic-resistant cells exist thanks to a random mutation. When the antibiotic is introduced into the culture, all sensitive bacteria will be killed, whereas resistant bacteria will continue to multiply by transferring this new feature, namely resistance to rifampicin, to subsequent generations. This example highlights the principles of genetic variation and natural selection. Rare DNA replication errors that occur randomly are the source of genetic variation. Depending on the characteristics of the environment—like the existence of rifampicin—bacterial populations with mutated genome are more resistant and thus more likely to survive.

In addition to these random mutations, bacterial cells can acquire genes from other bacteria, which can be transferred from one to another through the so-called horizontal gene transfer (HGT) or lateral gene transfer (LGT).[95] The behavior of *E. coli* is a typical example of this mechanism, too. In a culture of *E. coli*, some bacteria may be unable to produce the amino acid leucine because of a mutation and therefore cannot grow on a culture medium that does not contain this amino acid. Nevertheless, this particular strain has the ability to produce another amino acid, methionine. At the same time, some mutated cells in the same culture are able to produce leucine, but not methionine. If these two populations are found simultaneously in the same culture for some time, a new generation will emerge bearing a new feature; this strain will be able to produce both leucine and methionine. Thus, the genome of the bacteria of the new strain will include a normal gene for the production of leucine and another gene for the production of methionine. These exchanges and replacements of parts of their genomes allow bacteria to adapt to changes in the envi-

95 Ochman 2000; Dunning 2011; Keeling 2008; Gyles 2014.

ronment. The genes of one bacterium can be transferred to another in several ways, including transfer directly from cell to cell, during conjugation of two bacterial cells (bacterial mating).

Not all bacterial cells have the ability of conjugation and gene transfer. This ability stems from specific genes contained in bacterial plasmids that are independent of the bacterial chromosome. Apart from bacterial conjugation, some bacteria take up DNA molecules that are present in their environment after the death and degradation of other bacteria. These 'orphan' DNA fragments are incorporated into the bacterium's genome by recombination. This process is called transformation; the term derives from early observations when a bacterial strain appeared as if it transformed into another.

When a DNA fragment is introduced into a bacterium, either by conjugation or transformation, a sequence of processes begins. If the DNA fragment is a plasmid, it can be replicated independently and inherited to the offspring of this cell every time the cell is divided. If the DNA fragment cannot be replicated, it has to become part of the bacterial chromosome in order to be able to be passed on to daughter cells, otherwise it will be lost. The process of integrating DNA into the bacterial chromosome is called homologous recombination.

The genome of many bacteria contains DNA sequences called transposable elements or transposons, which are able to change their position within the chromosome. These transposons essentially lead to genetic variation since they accelerate the evolution of genomes. Transposable elements can also jump from the bacterial genome into plasmids and then be transferred to other cells by bacterial conjugation.

Finally, the transfer of genes among bacteria can also take place without contact, but through viruses (process of transduction). More specifically, the transfer of genes can be carried out by viruses that infect bacterial cells, the so-called bacteriophages the genome of which is integrated into the bacterial chromosome.

Genomic islands

Bacterial genomes often contain DNA 'islands' transferred through other species, the so called 'genomic islands' which can offer bacteria specific properties. Genomic islands (GIs) are clusters of genes within a chromosome that are horizontally transferred from other organisms. Based on the genetic properties of these genes, GIs can be categorized into: a) *pathogenicity islands* (PAIs), which encode various virulence factors; b) *metabolic islands*, which encode the new metabolic properties acquired by bacteria and allow them to adapt to a new environment; c) *secretion islands*, which encode the genes of secretion systems; and d) *antibiotic islands*, which encode antibiotic resistance genes.[96] It is believed that their formation contrib-

---

96 Che 2010.

utes to the differentiation and adaptation of microorganisms, which has a significant impact on the plasticity and evolution of the genome, as well as on the spread of antibiotic resistance genes in our time.[97]

Pathogenicity evolution

The process by which the structure and organization of a species' genetic information changes over time is known as genome evolution. Loss and acquisition of genes are changes that may rapidly and radically change the way a bacterium lives, to such an extent that their evolution is considered as taking 'quantum leaps'. The mechanisms of exchange of genetic material seem to be the primary forces that allow bacteria to genetically adapt to new environments and make them stand out from other microbial species. The process of pathogenicity evolution is a critical event in the life of a microbe.[98] The particular elements of genomic islands, 'pathogenicity islands' (PAIs), play an important role in this process. The concept of PAIs was initially introduced in the late 1980s, when scientists were investigating the virulence of UPEC (Uropathogenic isolates of *Escherichia coli*) at a genetic level.[99]

PAIs carry genes encoding one or more virulence factors which are absent from the genomes of non-pathogenic bacteria of the same or closely related species. Analyses have also shown that bacterial genomes consist of a mosaic genome architecture that comprises various portions of guanine and cytokine (G-C). The G-C content of genome sequencing is mostly homogenous (usually 70–80%) and distinct for each bacterial species. This stable 'genome-core' contains the genetic information required for the basic bacterial cell functions. On the other hand, the remaining genome (20–30%) carries unusual portions of foreign DNA with particular G-C content.[100] The G-C percentage content of PAIs is often different from the respective content of host organisms. The differences in G-C content between PAIs and core genomes have been identified in numerous bacteria, including *Y. pestis*.[101]

The evaluation of the change in G-C content is useful for identifying horizontally transferred genes, which is a global event across bacterial evolution.[102] In contrast to the theory of Charles Darwin, who suggested in 1859 that the evolution of species is a very slow process in small steps, the concept of PAIs acquisition by a microbial species allows it to make 'quantum leaps' as far as its genetic diversity and, consequently, its evolution are concerned. In some cases, the introduction of PAIs may lead to a

---

97 Juhas 2009.
98 Schmidt 2004.
99 Ho 2009; Novick 2016.
100 Hacker 2000; Gal-Mor 2006.
101 Dongsheng 2014.
102 Zhang 2014.

dramatic or even total change in a bacterium's phenotype or way of life, and convert it from a 'harmless' bacterium to a pathogen. For example, the ancestor of *Salmonella enterica* was a bacterium that was unable to invade the epithelial cells of an organism. However, the acquisition of fully functional PAIs created new generations of *Salmonella* with new properties and capabilities, signaling the species' transition to a more aggressive microbe that can now invade and adapt to the intracellular environment.[103] PAIs have been identified and studied in pathogenic strains of *E. coli*, *Salmonella* spp., *Shigella* spp., *Yersinia* spp., *Helicobacter pilori*, among others.

Another important property of PAIs is that they contain the genes related to the infectivity of the microorganism. According to the environment where bacteria live, the proteins encoded by genes associated with the infection can facilitate the adhesion to, and invasion of, eukaryotic cells by bacteria, toxin production, iron uptake from the host cell, and other properties. Moreover, larger quantities of virulence agents like toxins and type III and IV secretion systems were found in the genomes of pathogenic rather than non-pathogenic bacteria. On the other hand, larger quantities of other categories of infectious agents like motility, antiphagocytosis and iron (Fe^{+2}) intake, were found in the genomes of non-pathogenic rather than pathogenic bacteria.[104]

It is understood that—besides the constant rules by which bacterial cells copy their DNA and pass it on to the next generation—the bacterial genome is in a "fluid state" where mutations take place, while the genes and plasmids are transferred from cell to cell. Most genome modifications are harmful to the particular bacterium and thus disappear quickly from the population through the gradual death of mutant cells. Nevertheless, since bacteria are constantly in different environments with various forces of selection, the mechanisms that ensure genetic variation prove valuable since they offer the possibility of a new strain that is better adapted to the challenges posed to the survival of a bacterium. In conclusion, the spontaneous errors in DNA replication within a bacterial population are a constant source of genetic variation. When external conditions change, a strain that is better adapted to surviving in the new conditions will multiply and become the dominant population of this bacterium.

From "Innocent Pathogen" to "Natural-Born Killer": Evolution of *Yersinia pestis*

The basic—and at the same time the most specialized—functions of a microbe are regulated by genes. A gene is characterized as structural when it regulates the structure of a specific protein or a metabolic process of the microbe. Structural genes are

---

103 Zhang 2014.
104 Zhang 2014.

not in constant operation since their activity in the microbial cell is under control round-the-clock. Structural genes are bound to the bacterial chromosome along with other genes called operators, which regulate the activity of the former. This regulation is related either to the inactivation or to the activation of structural genes. The suppression of structural genes is carried out with the help of special protein substances produced by the cell called repressors. The production of repressors is in turn, bound to the control of other special genes (regulator genes). Such a gene sequence bound to the control of an operator constitutes a genetic unit called operon. *Y. pestis* contains operons comprising operators which control structural genes.

A microbe entails special genes, the activation of which increases its virulence. For example, as already mentioned, the basic factors of *Y. pestis'* virulence are the sophisticated type III secretion system and the Yersinia proteins (Yops).[105] At this point, specific genes particularly significant for the microbe's survival will be indicatively mentioned:

Haemin Storage System

This is a very important genomic system which binds the haemin chloride in which Fe^{2+} has been converted to Fe^{3+}.[106] Among the proteins, the synthesis of which is activated at low temperatures, are those expressed by the hms group of genes (*hmsH*, *hmsF*, *hmsR*).[107] In all *Y. pestis* strains with hms genes that have undergone mutation, the microbe is unable to colonize the oral cavity of the flea.

Fe2+ acquisition genes

In mammals, iron is bound to ferroproteins and hemoproteins. Upon their invasion, microorganisms must gain access to the iron of the host in order to use it for their multiplication and development.[108] This special mission is accomplished by the iron acquisition system under the combined expression of the genes *Hmu*, *Has*, *Ybt*, *Yfe* and *Yfu*.[109]

---

105 Califf 2015; Atkinson 2016.
106 Lillard 1997.
107 Darby 2002.
108 Ratledge 2000.
109 Carniel 2001.

caf gene cluster

As already mentioned, the antigen is considered to be a basic element of *Y. pestis'* virulence. The operon of the antigen consists of a group of genes that comprise the *caf1*, *caf1M*, *caf1 A* and *caf1R* genes. The final secretion and increase in antigen concentration is highly dependent on two genes, that is, *caf1M* and *caf1 A*.[110]

psaA gene (pH6 antigen)

For *Y. pestis* to be fully virulent, the *psaA* gene is necessary. It is expressed *in vitro* when pH is between 5–6.7 at 35–41°C.[111] The production of the pH6 antigen generally increases with temperature rise and the gradual drop of pH.[112]

LcrF RNA thermometer

RNA thermometers regulate gene expression in response to temperature fluctuations, enabling the stimulation of 'silent' genes upon the invasion of the cell by the microorganism. Together with a special protein, *LcrF* RNA thermometer activates the synthesis of the crucial *LcrF*, which enhances the microorganism's virulence. The sequence of RNA thermometer is similar in all *Y. pestis* strains that are pathogenic to humans.[113]

Ail

The ail locus encodes a thermally regulated surface-associated protein (expressed at 37 °C). Its mission is to allow the Yops proteins to enter the host cell faster.[114]

As the few examples presented above illustrate, the microbe's genes carry out highly specialized functions which are necessary for its survival. Where exactly do these highly specialized *Y. pestis* genes come from? How did the microbe acquire the necessary properties in order to survive and spread? Among the most popular methods to examine a species' origin is phylogenetics, which is defined as the study of evolutionary relationships among biological entities (species, individuals, or genes). The phylogenetic study of microbial strains from different regions largely reveals the relationship between species. Moreover, the number of differences among

---

110 Zavialov 2003.
111 Price 1995.
112 Konkel 2000.
113 Böhme 2012; Schwiesow 2015.
114 Atkinson 2016.

homologous DNA sequences can be used in phylogenetic analysis as a measure of genetic distance between species.[115]

Therefore, according to phylogenetic analysis, it appears that *Y. pestis* is most closely related to *Y. pseudotuberculosis*, which is believed to have evolved from *Y. enterocolitica* sometime between 40 and 190 million years ago. It should be noted that the dates stemming from the model calculations cannot be regarded as definitive.

The genus *Yersinia* constitutes a useful model for the study of the evolution of bacterial pathogens. Existing data indicate that the life of *Y. pestis* is associated with the gastroenteropathogen *Y. pseudotuberculosis*. As far as *Y. pseudotuberculosis* is concerned, it appears to have originated from another gastroenteropathogen, namely *Y. enterocolitica*. These two species cause a foodborne infection known as yersinosis, with gastroenteric symptoms such as diarrhea, vomiting, abdominal pain and fever.[116]

Y. pestis is often divided into three biovars according to the ability of the bacterium to ferment glycerol and reduce nitrate. Thus, the following biovars are distinguished: *Antiqua (*ferment glycerol and reduce nitrate), *Medievalis* (cannot reduce nitrate) and *Orientalis* (unable to ferment glycerol).[117]

From an evolutionary standpoint, *Y. pestis* is considered to have originated from *Y. pseudotuberculosis* (possibly serogroup O:1b), which leads to the conclusion that the plague microbe shares minimal common features with its ancestor. According to calculations that will be explained and discussed further on, this event seems to have taken place sometime between 2,000 and 6,000 years ago.[118] Moreover, the study of the two species' DNA reveals a DNA homology of 83%, which is greater than the species definition criterion defined at 70%.[119] According to genomic research, the genealogy of *Y. pestis* includes five major branches. It is assumed that the evolutionary process is associated with the effect of alternate cycles of enzootic and epizootic diseases in sylvatic plague foci.[120] The main branches of *Y. pestis* will be briefly mentioned here and further explained in the chapter on pandemics.

The study of *Y. pestis* is based on the concept of Single Nucleotide Polymorphism (SNPs), as the most common type of genetic variation. Each SNP constitutes a genetic variation in a DNA sequence when a simple nucleotide is mutated. SNPs are usually considered as adequately 'successful' point mutations that occurred in a significant percentage of a species population. The use of SNPs allowed us to reveal the first sequences of *Y. pestis* genomes, thus helping to compare the various strains of the same microbe as well as the microbe and other species. These SNPs enabled Mark Achtman et al. to define a basic three-branch population structure of *Y. pestis* in

115 Sallares 2007: 246.
116 Sabina 2011; Reuter 2014; Salomonsson 2016-2018: 13 – 18; Seecharran 2017.
117 Devignat 1951: 247 – 263.
118 Perry 1997; Achtman 1999 and 2004.
119 Bercovier 1984.
120 Cui 2016.

2004. Based on this structure, branches 1 and 2 originated from the ancestral branch (branch 0), with the ancestral status of the synonymous SNPs defined according to their status in the genome of *Y. pseudotuberculosis* (IP32953). New branches and sub-populations have been identified since, as the number of analyzed genomes is constantly rising.[121] The populations have been defined in line with the branch and the addition of the biovar abbreviation, for example ANT (*Antiqua*), PE (*Pestoides*), MED (*Medievalis*), ORI (*Orientalis*). The genomes of modern isolates disclosed the phylogenetic relationships between existing populations, while the ancient DNA studies provided data on the extinct lineages.[122] Further analysis of SNPs led to the construction of a more extensive phylogenetic tree comprising five branches: branch 0 (modern representatives of pestoids and *Antiqua*), branch 1 (*Antiqua* and *Orientalis* modern representatives), branch 2 (*Medievalis* and *Antiqua* modern representatives), branches 3 and 4 (modern representatives of *Antiqua*).[123]

Besides the relations between the strains of *Y. pestis*, Phylogenetics also revealed some properties related to its ancestor, *Y. pseudotuberculosis*. Despite the relation between the two species, particular differences have been identified mainly with regard to the virulence mechanisms. These differences reflect the very distinct nature of the infections, since the gastroenteropathogens *Yersiniae* cause self-limiting chronic infections, whereas plague leads to an acute disease that kills susceptible hosts within a short period of time. The speed of the plague disease possibly indicates that the evolution of the microbe did not pay particular attention to combating the defense mechanisms of T-cells in mammals. *Y. pestis* appears to be surpassing the problem of T-cells, while the responses of the latter against the disease are in any case belated.[124]

Unlike *Y. pestis*, the species *Y. enterocolitica* and *Y. pseudotuberculosis* continue to cause a significant number of gastrointestinal infections as a result of ingestion of infected meat products. As mentioned above, pathogenicity islands play an important role in a microbe's evolution. The full virulence of *Yersinia* species is founded in the high-pathogenicity island (HPI), which is present in *Y. pestis* and the increased-virulence serotypes of *Y. pseudotuberculosis* and *Y. enterocolitica*. Analyses of HPI in the three pathogenic species indicated that the locus has a G-C content of 60%, which is higher than the average G-C content of 46 to 50% for the core genome of *Yersinia* spp. The analysis of HPI in *Yersinia* spp. indicates so far that this structure constitutes an evolutionary event.[125]

However, the species that reflect the evolution of *Y. pestis* are characterized by more differences. For instance, the secretion of the signaling protein TNF-a by macrophages is more limited when the organism is infected by *Y. pestis* compared to se-

---

121 Achtman 2004: 17837–17842.
122 Wagner 2014: 319–26; Bos 2016: e12994.
123 Cui 2013: 577–582; Drancourt and Raoult 2016: 911–915.
124 Smiley 2008: 256–271.
125 Schmidt and Hensel 2004: 14–56.

cretion upon infection with *Y. pseudotuberculosis*. This simply means that *Y. pestis* has become better adapted to suppressing the antibacterial function of a macrophage in comparison to its ancestor, *Y. pseudotuberculosis*.[126]

But what exactly triggered this transformation of some individual strain of *Y. pseudotuberculosis* until it eventually differentiated itself from *Y. pestis*? *Y. pestis* and *Y. pseudotuberculosis* are still listed as different species due to the differences in their virulence and transmission route. A lot of genes which are essential for the enteropathogenicity of *Y. pseudotuberculosis* have been inactivated. This event has served as an important advantage for the development of a new life cycle of the plague bacterium.[127]

It is believed that two independent new plasmids that the bacterium acquired gave it a series of new characteristics and led to the creation of a new species. A necessary precondition for the pathogenic process of both microbes is the existence of the pCD1 plasmid, which encodes the type III secretion system mentioned above. The comparison of the two microbes shows that—besides the acquisition of the new plasmids (pCP1 and pMT1)—more than 300 genes are not functional in *Y. pestis*, that is, 13% of total genes of *Y. pseudotuberculosis*.[128] Moreover, as far as insertion sequence (IS) elements (that is, transposable nucleotide sequences which differ from those of the remaining DNA) are concerned, it seems that *Y. pseudotuberculosis* has comprised much fewer in relation to *Y. pestis* species since their evolution split up. The rearrangement of IS elements and massive gene inactivation in *Y. pestis* played a critical role in the evolution of the new microbe, and determined its aggressive behavior.[129] The process of gene inactivation and loss is a common phenomenon during the evolution of bacterial pathogens. This event of genome decay occurs when a gene or gene cluster is no longer useful for the survival of a microbe, or when a microbe attempts to adapt to a new environment.[130] A potential evolutionary scenario which has been proposed suggests that *Y. pseudotuberculosis* gradually acquired the necessary genetic determinants—such as the two new pMT1 and pPCP1 plasmids—from the resident gut microbiota during co-colonization of the rodent gastrointestinal tract or the flea midgut, and occasionally causes fatal septicemia in animals.[131]

All these processes, however, apparently required plenty of time so as to evolve and develop. We should also take into consideration the fact that the evolution process is influenced by factors like environmental conditions and mutations. To mention an example, although mutations occur at a constant rate, in times of epidemics this

---

126 Bi 2012.
127 Rasmussen 2015.
128 Parkhill 2001; Chain 2004.
129 Chain 2004.
130 Cerdeno-Tarraga 2004.
131 Achtman 1999; Reuter 2014.

rate can either remain unaltered or accelerate.[132] In order to better understand this property of microbes, it will suffice to think of the new vaccines offered by the health authorities every year in response to the new mutations of the influenza viruses or to the adaptive changes of the Ebola virus during epidemic outbreaks.[133]

The genetic losses of *Y. pestis* may be related to the strict epidemiological cycle of the microbe, which started to drift away from that of *Y. pseudotuberculosis*.[134] After all, the genetic information that *Y. pestis* lost was no longer useful and was mainly related to the enteric life of *Y. pseudotuberculosis*. The medical press contains more and more studies that compare the genomes of *Y. pseudotuberculosis* and *Y. pestis*, to discover more differences that could help comprehend the evolution of microbes. These studies show that numerous genes are common in the strains of *Y. pestis*, but differ in *Y. pseudotuberculosis*. Indicatively, the comparison of three *Y. pestis* strains (CO92, KIM10, and 91001) with the genome of *Y. pseudotuberculosis* revealed more than 2,500 common genes in the three strains, which are nevertheless different in *Y. pseudotuberculosis*. The differences between the two species, which are also responsible for the differentiation of *Y. pestis*, are identified in various aspects of the microbe's nature. For instance, the new characteristics acquired by *Y. pestis* concern properties related to cell cycle control and motility, cell membrane biogenesis, extracellular structure, signal transduction mechanisms, metabolism energy production, inorganic ion/ amino acid/ nucleotide/ co-enzymes/ lipid transport and metabolism, among others.[135]

As far as the evolutionary change in the aggressiveness of *Y. pestis* is concerned, it seems that it influences the defense of the organism by affecting the basic functions of macrophages (phagocytosis, secretion of TNF-a and nitric oxide and other functions), which is not the case when *Y. pseudotuberculosis* invades the organism.[136] In other words, the strategy of the newly-emerging *Y. pestis* can easily be juxtaposed with the main principles of a well-designed military attack plan: breach in the defensive arrangement and establishment of a beachhead, interruption of opponent's communication/command and control functions, and paralysis of defense. This exact strategy has been employed by one of the most deadly microbes in the history of humanity.

---

132 Weil 2018; Nogales 2018; Möller 2018.
133 Dietzel 2017; Davidson 2018.
134 Pouillot 2008.
135 Gu 2007.
136 Bi 2012.

The Use of Polymerase Chain Reaction (PCR) in Molecular Studies

In recent years, numerous studies and books of medico-historical content have used data stemming from the laboratory application of molecular methods. In the daily laboratory diagnosis of patients, these molecular methods are governed by strict protocols, while their reliability is controlled by two basic statistical rates, namely the sensitivity rate (true positive result) and specificity rate (true negative result). The molecular methods of identifying pathogens are now being employed in the fields of paleomicrobiology and paleopathology. Experimental data show that the microbial DNA in a host can survive for up to 20,000 years, whereas in permafrost specimens it is maintained for more than a million years. The DNA of potential pathogens can be found in bone marrow and mummified tissues. Given that the endodontium (pulp of the tooth)—as a highly vascularized tissue—can contain microorganisms circulating in the blood, it is considered an adequate sampling material for the detection of potential pathogenic microorganisms. The DNA traced in specimens of the distant past is generally fragmented and corresponds to a small number of base pairs (bp). Moreover, the risk of contamination by extraneous DNA is always visible in specimens of the past, and poses a significant challenge when applying molecular techniques. Contamination can be identified through either microorganisms at the burial site where the skeletal remains were discovered, or microorganisms found in the laboratory during the processing of the specimen.

The most typical method mentioned in available sources is Polymerase Chain Reaction (PCR). The applications of PCR are currently related to a wide range of fields like Clinical Microbiology, Forensic Medicine, Genetic testing, population analysis, and cloning, for example. Considering that many variations of PCR have been discovered so far, only certain general elements pertaining to the main principles of this method shall be explored here. PCR is the process of amplifying a DNA sequence through replication *in vitro*. In other words, it is a simple way of replicating a genetic segment in order to allow its further study using various other methods. Without going into specialized technical aspects of this method, the basic elements of PCR reaction are the following:

DNA template

Comprises the particular DNA sequence that we want to copy.

DNA polymerase

DNA polymerase is an enzyme that is present in all organisms and participates in DNA replication. It cannot synthesize a new molecule on its own, but it is able to replicate an existing molecule that is used as a template.

Primers

Primers are oligonucleotides that define the DNA amount to be amplified.

Buffer and Magnesium ion (Mg^{2+})

Buffer preserves the ideal conditions for realizing the reaction. Mg^{2+} ions are necessary as a divalent of DNA polymerase.

Deoxynucleoside triphosphate (dNTPs)

These are the structural molecules used for synthesizing the new DNA chain.

The reaction is carried out in three recurring stages: i) denaturation (the two DNA strands are separated by heating at 94–95 °C), ii) annealing (drop of temperature to 55–65 °C so as to allow the primers to base pair with complimentary regions of the DNA template), and iii) extension (temperature rise to 72° C, polymerase elongates primers introducing dNTPs and using complementary DNA sequence as a template). The rate of synthesis of the new chain amounts to 1,000 base pair/min. The aforementioned stages are repeated 30 to 40 times (cycles). The process is carried out in a device called PCR Thermal cycler, which can regulate the temperature and duration of each stage.

More molecular methods are expected to be implemented in the coming years in the field of ancient DNA, such as next-generation sequencing technologies or DNA microarrays, which will possibly offer more insights into pathogens' phylogenetic evolution while further enriching the databanks of ancient-DNA.[137]

---

137 Selected sources for further reading on ancient DNA and PCR: Drancourt 1998; Raoult 2000; Willersley 2005; McCormick 2007; Drancourt 2008; La 2008; Devault 2014; Bos 2015; Vogler 2016; Llamas 2017.

Chapter 3
Aggravating Factors of Plague

Epidemiological Aspects of Plague

The word 'epidemiology' derives from the combination of the Greek preposition ἐπί (*epi*, meaning 'on' or 'upon'), and the Greek words δῆμος (*dêmos*, people) and λόγος (*logos*, with a broad meaning, to be understood here as referring to study). In other words, epidemiology is the study of what befalls a population. According to a most technical, accurate definition, epidemiology is the study of the distribution and determinants of health-related states or events in specified populations, and the application of this study to the control of health problems.[138] Until the middle of the twentieth century, epidemiology was focused on the epidemics of communicable diseases. After World War II, the field expanded and comprised non-communicable diseases, environmental, and occupational health. During the twenty-first century, the latest developments in the evolution of this scientific field have been termed as meta-epidemiology. The concept of meta-epidemiology takes into consideration the methodological limitations of the systematic review of intervention trials.[139]

Being a quantitative discipline, epidemiology employs probability and statistics research methods. Through causal reasoning it develops and tests hypotheses. Epidemiology examines cases or events, and describes them in terms of time, place, and affected individuals, calculating rates, and comparing them over time or among different groups of people.[140]

Epidemiology primarily focuses on the study of infectious diseases caused by infectious agents (or by their toxic products). These agents can be transmitted to humans either directly, via another human or an animal, or indirectly by a vector or particular objects of the inanimate environment. The diseases that spread from human to human without the intervention of a transmitter are also called contagious. For example, although malaria is infectious, it is not contagious, as the disease transmits via mosquitoes. On the other hand, measles is both an infectious and contagious disease, since it transmits directly from person to person. The epidemics caused by infectious agents can last from a few hours (such as food poisoning) to several decades, with cholera being the most typical instance of the latter case. Indicatively, the Haitian Ministry of Health estimates that the cholera epidemic in Haiti following the disastrous earthquake of 2010 will not cease before 2022.[141] The area where such epidemics are prevalent may range from a closed group or small area (for instance a

---

138 Last 2001: 61; Centers for Disease Control and Prevention 2012: 2.
139 Bae 2014.
140 Centers for Disease Control and Prevention 2012: 2.
141 Guillaume 2018.

https://doi.org/10.1515/9783110613636-006

camp, prison, or nursing home) to a whole continent and beyond. The so called epidemic outbreaks refer to epidemics of short duration (of a few hours, days or weeks), and are usually restricted to the local level, affecting specific neighborhoods or cities. When an area is continually affected by sporadic outbreaks of an infectious disease which are characterized by a certain frequency, the specific disease is characterized as endemic—or as enzootic when it affects animals.[142] An endemic disease may lead to an epidemic outbreak under specific circumstances. The major causes are: a) the sharp increase in insect-vectors of a disease due to ecological changes; b) the increase in the proportion of immune and non-immune individuals as a result of mass migration from epidemic areas or of inadequate inoculation programmes due to the negligence of health authorities; c) the existence of antibiotic-resistant strains of the microorganisms. When an epidemic (or epizootic) spreads over a wide geographical area covering one or more continents, it becomes a pandemic. Even in this case, however, no quantitative measure exists establishing when an epidemic becomes a pandemic.

Time is one of the most important parameters of an epidemic. A rough classification of time distinguishes between short-term fluctuations, temporal clustering, and cyclic fluctuations. In the case of short-term fluctuations, many susceptible individuals are simultaneously exposed to the same infectious agent. In this case, the duration of the epidemic is proportional to the incubation period of the relevant microorganism. A typical case is food poisoning in camps and schools, for example, with the simultaneous exposure of many individuals to the microbiological agent. Temporal clustering refers to cases where a set of individuals are exposed to the same infectious agent, but at different times. These times are significantly apart from each other, and the outbreaks occur with a corresponding delay, making it possible for their epidemiological correlation to go unnoticed. But if these times are close to each other, the outbreaks are characterized by temporal clustering, which is associated with the time when the individuals were exposed to the infectious agent. Finally, given that some infectious diseases exhibit seasonal fluctuations, certain epidemics are similarly characterized by seasonal fluctuations.

Regardless of their type, however, all epidemics are characterized by the same time phases, although there are frequent variations of the typical model. The time between exposure to an infectious agent and the appearance of the first symptoms is called *incubation period*. After the active exposure to the infectious agent follows the *latent period*, during which the infectious agent is not discharged into the environment. The latent period does not necessarily coincide with the incubation period. The process of discharge of the infectious agent into the environment is followed by an infectious period, which lasts as long as the discharge period. It should be noted that the infectious period is not consistent with the period when the clinical symptoms appear but rather precedes it. Therefore, the isolation of patients whose clinical

---

142 Riley 2018; Riley 2019.

symptoms are already apparent is not in itself sufficient to contain an epidemic, because new individuals are becoming infected (infectious period) during the period of patients' isolation.[143]

A reader can easily comprehend the rationale behind the basics of epidemiologic research in relation to the methodology of the popular *Crime Scene Investigation* TV-series (the *5 W's:* what, who, where, when, why). More specifically, as far as epidemiologists are concerned, the *5 W's* entail: case definition, person, place, time, and cause/risk factors/models of transmission.[144]

In the case of historic plague outbreaks, descriptive epidemiology examines the time, place and individual. The most important aspect of epidemiology is that no health event occurs as an isolated or random event, but rather as the result of complex interactions. At the core of these interactions, a group of risk factors can be identified. In epidemiology, a risk factor is a variable associated with an increased risk of disease or infection. A risk factor is defined as any attribute, characteristic or exposure of an individual that increases the likelihood of developing a disease or injury.[145]

Given that one of the basic goals of epidemiology is the identification of the causation of a health event, a number of models have been proposed. The simplest model for infectious disease outbreaks is the epidemiologic triad (or epidemiologic triangle). The triad consists of an external agent, a susceptible host, and an environment that brings them together. The onset of a disease or an epidemic requires different interactions between these three elements. External agents comprise microorganisms, namely bacteria, viruses and parasites. The main feature of infectious agents is their ability to cause a disease. Based on their nature, they are characterized according to their infectivity, pathogenicity, and virulence. *Infectivity* refers to the proportion of exposed individuals who become infected. *Pathogenicity* refers to the proportion of infected individuals who develop a clinically apparent disease. *Virulence* refers to the proportion of clinically apparent cases that are severe or fatal.[146] The invasion of infectious agents into the human organism is followed by pathological changes. The disease can range from mild to severe or fatal. This range is called *spectrum of disease*. The disease process can end in recovery, disability, or death of the patient.[147]

Host refers to the human population and includes factors which can contribute to the exposure and susceptibility to any infectious agent. The susceptibility of the host to the microbial agents can be influenced by factors like nutritional and immunologic status.[148] The concept of collective immunity is basic for the protection of a

---

143 Gould 2009.
144 Centers for Disease Control and Prevention 2012: 31.
145 Centers for Disease Control and Prevention 2012: 52.
146 Centers for Disease Control and Prevention 2012: 61.
147 Centers for Disease Control and Prevention 2012: 60.
148 Centers for Disease Control and Prevention 2012: 53.

community. The necessary degree of herd immunity to prevent or interrupt an outbreak varies by disease. If an adequate number of individuals in a population are resistant to an agent, those who are susceptible will be protected by the resistant majority.[149]

Environment, the third element of the triad, includes climatic and biologic factors. These factors affect the agent and the host's exposure. They include socioeconomic parameters like overpopulation, sanitation, and the level of the health services. The emergence of infectious diseases reflects changes in human ecology such as migration from rural to urban areas or long-distance mobility and trade.[150]

Apparently, the basic principle of the epidemiologic triad is timeless, and can be applied to human populations across the centuries. The case of plague in Byzantium is no exception and does not deviate from this general principle. It is also understood that, in the specific case of Byzantium, plague was inextricably linked with the human populations and the environment. In this context, we ought to examine the factors that contribute to the outbreak of an epidemic in the long run.

The Byzantines as Hosts: Geography and Demography

The word 'Byzantium' refers to the Medieval history of the Balkan region and the wider southeastern Mediterranean basin. In the political turmoil of the third century A.D., Emperor Constantine the Great (c. 272–337) decided—in the early fourth century A.D.—to transfer the administrative center of the Roman Empire from Rome to an old colony of the ancient Greek city-state of Megara in the Propontis: the city of Byzantium. This process was sealed by Emperor Theodosius I (imp. 379–395) on his deathbed in the late fourth century A.D., when the *Imperium Romanum* was divided into *Pars occidentalis* and *Pars orientalis* (western and eastern halves of the Empire), announcing the end of the Roman world—*Romanus Orbis*—and the gradual transition to the notion of a 'Byzantine Ecumene'.[151]

The constant opposition and rivalry between the eastern and western parts of the Roman Empire, coupled with the frequent incursions by *barbarian tribes* did not immediately allow for the definition of an exact boundary between the two parts.[152] It was only with an agreement of 437 and the concession of the region of Sirmium (in modern Serbia) to the Eastern Empire, that the European border of Byzantium was definitively determined, with the inclusion of the largest and most significant part of the Peninsula of Haemus (the Balkan Peninsula).[153] The Byzantine *mare nostrum* of the fifth century extended from the Black Sea to the northeast to the shores of

---

149 Centers for Disease Control and Prevention 2012: 68.
150 Weiss 2004.
151 Ostrogorsky 1968: 22–50.
152 Savvides 1983: 13–21.
153 Svoronos 1978.

North Africa to the south. At its peak, the Byzantine Empire stretched northwest to modern Croatia, and its northern borders covered part of modern Croatia, while, in the northeast, it reached the Danube Delta, the Crimean Peninsula, and the southern part of modern Georgia. In the West, in some periods of its history, Byzantium succeeded in controlling Central and southern Italy (and some parts of the Northeast Apennine Peninsula), Corsica, Sardinia, and the Balearic islands up to the Straits of Gibraltar, whereas, in the South, it retained—for some centuries—Egypt and the coasts of North Africa up to modern Morocco. The easternmost boundary of the Empire reached Lake Van in modern Turkey.[154]

Although the Byzantine Empire was characterized by a heterogeneous, multinational, multilingual, and multireligious identity through the different stages of its history, its administrative system of centralized monarchy remained unchanged, and rested on a single state, religion-church and administration apparatus, which facilitated the preservation of this heterogeneous entity. A simple count of the ethnicities in the Empire is enough to demonstrate its multiethnic character. Greeks, Armenians, Egyptians, Syrians, Slavic tribes and remnants of Asia Minor populations (Cappadocians, Lycaens, Isaurians, Phrygians), Balkan ethnicities (Illyrians, Scythians, Thracians, Moesians), Lazes and Abasgians (as allied peoples), were the first to make up the population of the Empire in the fourth century.[155]

An effort to quantify the population of the Byzantine Empire at different time periods would not only be difficult, but rather utopian—leaving us only with speculations. Available figures are mainly related to the major urban centers and must be considered regarded as provisory. During the fourth century, the population of the Eastern Roman Empire was generally larger than that of its western counterpart.[156] It is speculated that the population of the Empire was approximately 26 million people during the fourth century, of which five million lived in the Balkan Peninsula.[157]

Excavations have revealed intense residential occupation and densely populated areas in the Peninsula of Haemus. However, frequent barbaric raids on the Danube led to the destruction of large urban centers and a sharp drop in the population, especially in the countryside.[158] The emperors Theodosius II (imp. 408–450), Marcian (450–457), Heraclius (610–641) and his successors, implemented settlement policies and tolerated the influx of barbarian tribes, allowing them to settle in in the territories of Byzantium (upon recognition of the Byzantine imperial supremacy). These policies gradually mitigated the demographic problem of the Empire.[159]

The raids on the northern frontiers had been a constant preoccupation for the emperors of the Early Byzantine period. Their impact on the cities of Upper Moesia

---

154 Savvides 1983: 13–21.
155 Savvides 1983: 13–21.
156 Svoronos 1978: 278–292.
157 Charanis 1967: 446.
158 Eadie 1982.
159 Savvides 1983: 13–21.

(*Moesia Superior*), *Dacia Ripensis*, *Dacia Mediterranea*, Lower Moesia (*Moesia Inferior*), and Scythia was terrifying. The destruction of Naissus (modern Niš in Serbia) and the devastation of the wider region of *Regio Sirmiensis* had a great psychological impact on the local populations.[160] The recount of the Byzantine general Priscus (d. 613 A.D.) regarding the situation in Naissus is indicative:[161]:

> Ἀφικόμενοι δὲ εἰς Ναϊσσὸν ἔρημον μὲν εὕρομεν ἀνθρώπων τὴν πόλιν, ὡς ὑπὸ τῶν πολεμίων ἀνατραπεῖσαν, ἐν δὲ τοῖς ἱεροῖς καταλύμασι τῶν ὑπὸ τῶν νόσων κατεχομένων τινὲς ἐτύγχανον ὄντες.

> Arriving at Naissus, we found the city empty of its population, as if they had been expelled by invaders, while some of those affected by the diseases happened to be in the sacred places.

Even in the time of Justinian in the sixth century, the situation in the Danube region could be described as hopeless, "it is surrounded by extensive and largely deserted areas" (χώραν δὲ περιβάλλει πολλήν, ἐκ μὲν τοῦ ἐπὶ πλεῖστον παντελῶς ἔρημον).[162] The raids of the Goths and Huns on the Balkans essentially paved the way for the raiders of the coming centuries. That period was characterized by the extensive destruction of cities, fortresses, and villages, the devastation of arable land, and, most significantly, by the destabilization of the Byzantine military-defense system, which led to the reduction of the local population and the suppression of commercial activities.[163]

According to Emperor Constantine VII Porphyrogenitos (imp. 913–959), in the time of Heraclius (that is, in the seventh century) and his successors, new tribes such as the Croats and the Serbs entered massively and settled in the territories of the Empire.[164] The situation gradually normalized as these tribes accepted the imperial authority and integrated into the politico-economic (offices and tax obligations), religious (Christianization), and military apparatus of Byzantium.

In the East, the Byzantine victories during the fourth and fifth centuries against the Persians, the creation of a series of frontier formations or subjugated tribes in Asia, and alliances with the Ethiopian peoples in Africa contributed to a gradual increase of population.[165] At the time of the *Plague of Justinian* (sixth century) and until the seventh century, the population is estimated at approximately 30 million.[166]

At the time of the Arab conquest (that is, in the seventh century), it is estimated that the population of Egypt, Syria, Palestine, and the region of the West Bank, which ceased to belong to the Empire as these provinces were never recovered,

---

160 Dagron 1984.

161 Svoronos 1978: 278–292.

162 Procopius, *De aedificiis*, IV.5.9 (ed. Haury 1913: 125.10–11).

163 Patoura-Spanou 2008: 51.

164 Constantine Porphyrogenitus, *De Administrando Imperio*, 29.53–78 (eds. Moravcsik and Jenkins 1967: 124–126); Maksimovic 1994: 9–12.

165 Savvides 1983: 13–21.

166 Charanis 1967: 446.

amounted to 16 million.[167] From the late eighth and early ninth century onwards, the demographic evolution of the Empire seems to have regained its earlier pace, as a result of the settlement policy, forced movements of populations, and the settlement of Slav-Bulgarian and Armenian-Syrian populations.[168] Considering that the population of the Byzantine Empire was indeed 19 million in the ninth century and 20 million in the eleventh century, this rise is impressive and amounts to a rate of 88%.[169] During the eleventh century, the population of Byzantium might have reached 20 million, but it gradually declined following the decadence of the Empire as a whole. Notably, at the time of Michael VIII Palaiologos (that is, in the thirteenth century), the population of the Empire might have been around four million.[170]

The Byzantine Empire was a dynamic population system: the massive influx of new tribes and their voluntary or forced acceptance of the Byzantine reality gave a cosmopolitan, inclusive, and multicultural character to the Empire. Its urban centers were attractive poles and the final destination for many internal movements. Nevertheless, Byzantine cities were often unable to host new inhabitants, and the subsequent overpopulation led to forced urban planning and environmental interventions that affected the health of their residents, opening the door to epidemic diseases.

The Byzantine City as a Risk Factor

The study of Byzantine cities is generally difficult, and individual time periods in their evolutionary course can only be approximately distinguished. These periods can be summarized as follows:[171]

i)   Third to early sixth century: during this period, the cities were still confined within the old Roman walls. Many were bound to be destroyed by barbarian raids, while the new ones that emerged were built on the basis of a new residential model.

ii) Mid-sixth to eighth century: Byzantine cities were completely differentiated from the strict Roman urban planning system, their size increased, and they adapted to the geomorphology of the environment, while the first phenomena of urbanization appeared.

iii) Eighth century to 961 A.D.: new problems emerged due to urbanization and the destruction of cities by raids. This period is considered critical since the cities literally struggled to survive in the general setting of the so-called 'Byzantine Dark Ages'.

---

167 Savvides 1983: 13–21.
168 Svoronos 1978: 278–292.
169 Svoronos 1978: 278–292.
170 Savvides 1983: 13–21.
171 Bouras 1981.

iv) 961–1071 A.D.: these 100 years have been defined by historians as the era of 're-birth' of the Byzantine cities: their population increased, urban life was normalized, and fortifications extended to protect the settlements outside cities.

v) 1071–1204: this period was marked by the decadence of the cities in Asia Minor, and the flourishing of the cities in mainland Greece.

vi) 1204–1261: the Frankish conquest of the imperial territory concluded with the recovery of the cities in Asia Minor and the reconquest of Constantinople under the rule of the Laskarid dynasty. At the same time, the cities of the Empire of Trebizond and the Despotate of Epirus prospered.

vii) 1261–1453: the weakening of the central administration allowed for more liberty of some cities in the management of their urban economies, which resulted in their final revival before they were gradually taken over by the Ottoman Turks.

The early Byzantine city was essentially the continuation of the former cities of the Late Roman Antiquity. The decline of this kind of cities during the sixth century can be considered as the result of their transition to the model of the medieval Byzantine city. Old Roman cities were built on a square or rectangular shape in line with the rigid system of the two central axes, *cardo maximus* (North-South axis) and *decumanus maximus* (East-West axis), that is, in the way the camps of Roman legions were built.[172] These cities had a small size, as they were necessarily limited by their walls and conformed to a rigid and strict Roman residential plan. During the early Byzantine centuries, this practice was gradually replaced by a less stringent model adapted to the geomorphology of the terrain.

At the time of the *Plague of Justinian* in the sixth century it is estimated that the Empire had approximately 1,500 large or smaller cities.[173] In the context of the reorganization of the Empire by Justinian (the so-called *renovatio imperii*), new cities were built, and the existing cities were reconstructed with a focus on their defense. This does not imply that the Byzantines did not provide for the healthcare of their citizens. A typical example was the village of Vizana, mentioned by the Byzantine historian Procopius (500–565 A.D.) in his work *De aedeficiis* (*On Buildings*, Περὶ κτισμάτων). Procopius describes a place in Armenia with swamps and an unhealthy climate, the inhabitants of which suffer from an infectious disease, probably related to malaria.[174] Due to the adverse climate, diseases, and frequent raids, the Emperor Justinian decided to build a mountain city located three kilometers away from Vizana, in the area of Tzumina, where the fresh air improved the well-being of the citizens who settled in this new city. This concept of clean air provided the basis for the regulations of the operations of limestone mines in the capital city of Propontis and the port of Julian (*Portus Novus*). These regulations stipulated that limestone mines had

---

172 Kordosis 1996: 219.
173 Mango 1980: 60–88.
174 Procopius, *De aedificiis*, III.5.13 (ed. Haury 1913: 95.9–14).

to be located at least 100 cubits (45,7 meters) away from every building on each downwind, because the smoke coming from the mines was very intense and poisonous (Περὶ φούρνου τῆς ἀσβέστου—*Edictum de Fornace Calcis*).[175]

While the Early Byzantine Empire was a set of cities, the Middle- and Late-Byzantine Worlds were rather a set of castles.[176] Pergamon, for example, was transformed into a small fortress after the Arab invasions in 664, 716–717 and 737.[177] The remnants of Byzantine Corinth, Athens, Pergamon, and Ephesus suggest an image of unregulated building or reconstruction for the general defense of the city.[178] Middle- and Late-Byzantine settlements usually enclosed houses poorly built out of cheap materials (excluding the nobles' residences), while their streets were unregulated and labyrinthine, and often led to dead ends, making the passage of wagons impossible.[179]

The Empire comprised many more major urban centers besides Constantinople: Antioch, Thessaloniki, Alexandria, Miletus, Sirmium, Mesembria (modern Nesebar), Adrianople (modern Edirne), and Singidunum (modern Belgrade). The various sources about the cities of Asia Minor mention numerous cities that were large and populous thanks to the particular economic and military importance of the region.[180] Judging from its description by its inhabitant Ioannes Kaminiates, 10th-century Thessaloniki might have been a typical case of such a city with the capability of extending its residential quarters within its walls before its conquest by the Saracens in 904 [181]:

Εἴπομεν ὡς εὐρεῖά τίς ἐστι καὶ μεγάλη ἡ πόλις καὶ τῷ περιέχοντι πολὺ τὸν διὰ μέσου χῶρον ἐναποκλείσασα.

We said that the city is large and spacious, and all that it contains comprises a sizeable area.

It is estimated that it actually encompassed a considerable area within its walls: 1,750 meters from the east to the west, and 2,100 meters from the north to the south.[182]

Antioch, the third largest city of the Empire, occupied an area of 650 hectares in the sixth century. Daras, a very significant fortress-city on the eastern border of Mesopotamia, stretched over an area of 1,000 by 750 meters, while Laodicea, with its size of a mere 220 hectares, seemed large compared to the other cities of the Thracesian Theme of Asia Minor.[183]

---

175 Constantine Harmenopoulos, *Hexabiblos* 4.2.17 (ed. Heimbach 1851: 246); Tourptsoglou-Stefanidou 1998: 140–167.
176 Mango 1980: 60–88.
177 Vlyssidou 1998: 221.
178 Bouras 1981.
179 Runciman 1980: 87–97.
180 Lounghis 1998.
181 Johannes Caminiates, *De expugnatione Thessalonicae,* 8 (27–28) with the description of the city (ed. Böhlig 1973: 9).
182 Mango 1980: 60–88.
183 Mango 1980: 60–88.

The actual population of a city based on its size is difficult to calculate because of a series of imponderable factors such as their streets, squares, and gardens, or fortification specificities (for example, the distance of settlements from the inner walls). Moreover, any information on the population of a city prior to the year of its destruction cannot be used as a *de facto* model for the period after its reconstruction. The widespread destruction of urban centers radically changed the planning and population of cities, since the fear of a possible return of the barbarians led many of the inhabitants to flee. The desolation of Athens after its terrible destruction by the Goths and Herules in 267 A.D. is a typical example.[184] The available data on the population of specific cities must be considered as approximations. Certain indicative figures will be mentioned, but not without reservation. According to the Antioch-born Father of the Church and Archbishop of Constantinople Ioannis Chrysostomos (349–407), the city of Antioch had 100,000 inhabitants in the fourth century, whereas between 120,000 and 500,000 people lived in Alexandria.[185] It is estimated that the population of Constantinople doubled between the fourth and the sixth century, from 200,000 to 400,000 inhabitants.[186] Indeed, during the mid-fourth century, the rhetor Libanius (314–394) characterized Constantinople as a densely populated city that devoured people, while, in the fifth century, Theodoritus described it as a universe populated by an ocean of human beings.[187] By the twelfth century, the population of Constantinople had risen to 500,000 or even 800,000.[188] According to another approach, at the time of the *Plague of Justinian*, the population of the city may have reached one million.[189] As far as Constantinople is concerned, legend has it that the Emperor Constantine the Great himself defined the limits of the new capital city of his Empire. During his reign, the city tripled in size compared to what it was during the reign of Septimius Severus (imp. 193–211), when the city covered an area of 200 hectares and accommodated 20,000 or 30,000 residents.[190]

The choice of Constantinople as the new capital of the Empire by Constantine was ideal, and its design was excellent. Naturally, in every city the streets form the main urban axes, whereas proper planning defines its future development. Similarly, Constantinople developed along specific main streets and according to a plan that envisaged its future expansion.[191] The central artery of the city was the so-called *Mese*, one branch of which connected the *Golden Gate* (*Porta Aurea*) with the Forum of Constantine and extended from the *Milion* to the *Chalke Gate*. The second branch

---

184 Gregory 1982.
185 Joannes Chrysostomus, *Homilia* LXXXV.al.LXXXVI (ed. Migne 1862: 762–763); Liebeschuetz 1971: 92–100.
186 As reported in Magdalino 1996: 57.
187 As mentioned by Concina 2003: 39–46.
188 As per Concina 2003: 39–46.
189 Andréades 1920.
190 Dagron 1974: 13–48.
191 Mango 1985: 41.

stretched westwards from the *Forum of Theodosius* to the *Church of the Holy Apostles* and ended in the *Gate of Charisius*. The *Mese* was paved with stone slabs, and other streets extended from it, such as two large vertical road axes.[192] Public buildings and private homes were built along these central arteries. The major layout and grid of the residential quarters were shaped like a normal reclining Y, orientated from the east to the west.

Shortly before the completion of construction works in the new capital, the first members of the Senate started to arrive and settled in with their families. The buildings for the upper class of public administrators have been identified by various names, such as μέγισται οἰκίαι *(megistai oikiai—largest houses)*, οἰκίαι *(houses)*, or *domus* – οἰκίαι *(domus—insulae)*.[193] Contemporary scholars disagree as to the exact nature of the *domus* – oikiai. Allegedly, in the early years of the foundation of Constantinople, a new type of 'folk' dwelling may have been established, which combined *domus* and *insulae*, although the character of the former may have been more family-oriented.[194] Apparently, until the early fifth century, the city was able to cope with the constantly increasing influx of new residents. At that point, however, it reached and exceeded its maximal capacity, and the ensuing demographic explosion caused numerous problems on a residential and sanitary level.

The new makeshift settlements expanded beyond the boundaries of the wall that was built during the reign of Constantine the Great. The residents did not feel safe until 413 A.D., however, when that newly built area was protected by the the Theodosian walls, which can be considered as a miracle of fortification.[195] Many poor constructions made of piles and reeds were also built on the side of the Bosporus and the *Golden Horn*.[196] Apart from splendid buildings, palaces, and churches, some of Constantinople's districts were certainly not reminiscent of the city's status as the capital of the Empire. Maybe the drafting of specific legislation of urban planning and public health indicates the magnitude of the residential unaccountability and/ or of the citizens' indifference. These impressively detailed laws define with great precision the required distances between public buildings and private homes, as well as between private homes themselves. One of the most interesting provisions involves the judicial payment, or fine, of the ἀερικόν *(aerikon-air)*, which historians of legal history believe to have been a fine on constructions without a building permit that deprived neighbors of their view to the sea.[197] It is worth mentioning at this point that the Byzantines paid great attention to the free access of citizens to the sea, the view of which was considered to be an entertainment. The regulations on buildings on the side of the sea walls might have been based on more than the citizens' enter-

---

192 Sinakos 2003: 50 – 53.
193 Sinakos 2003: 50 – 53.
194 Sinakos 2003: 50 – 53.
195 Mango 1985: 41.
196 Dagron 1974: 13 – 48.
197 Koukoules 1951: 249 – 336; Karpozilos 1989.

tainment, namely on practical issues. It was necessary for the sea breeze to reach the inner parts of the city in order to make its atmosphere more breathable, especially during the summer. The dense construction of tall buildings near the sea walls would have been an obstacle.

Many houses had more than one story, with various reports referring to five-story buildings which were inhabited mainly by people from the lower strata. The account of the situation in a multi-story building of Constantinople by the twelve-century writer Ioannis Tzetzes is both humorous and revealing. Tzetzes lived in a three-story house; in a letter to the senator Nikephoros Serblios, he describes in great detail the drama that he experienced every day due to the indifference of his priest neighbor.[198] The priest who lived on the upper floor, had many children and he used to throw the family's waste on the street, while a broken gutter led the dirty water straight into Tzetzes' room. In a poignant manner, Tzetzes likens the quantity of dirty water and urine that accumulate in the streets of the neighborhood to a river, on which even a ship could sail.[199]

Tzetzes' narrative is definitely exaggerated, but surely the indifference or economic inability of some citizens to repair the damages to their homes contributed to the gradual deterioration of health conditions in the city. Of course, no one could have ever seen a ship sail on the street of Tzetzes' neighborhood. However, one could see all kinds of scrap and waste on Byzantine urban streets, despite the fact that it was legally forbidden to dispose excrement, dead animals, and leather.[200] Odon de Deuil, the French historian and traveler of the twelfth century, described Constantinople as dirty and full of waste.[201] A similar situation was probably evident in the 'castle of the Franks', that is, the Genoese colony of Peran opposite Constantinople, which was founded in 1267 A.D. by permission of the emperor Michael VIII Palaiologos (imp. 1259–1282). Although Andronicus II Palaiologos (imp. 1282–1328) allotted another six hectares to the Genoese, and Peran reached a total of 37 hectares. According to the book *Rihla* (c.1355) of the Arabic traveler Ibn Battuta (1304–1368 or 1378) the colony was dirty, and streams of dirt and waste water run through its central market all the way to the busy harbor.[202] The residents of Byzantine cities also improvised new dumps, which were often located near the city walls.[203] Procopius' description of the Hierapolis lake (modern Pamukkale, Turkey)

---

198 Joannes Tzetzes, *Epistulae* 18 Τῷ μυστικῷ κυρῷ Νικηφόρῳ τῷ Σερβλίᾳ (ed. Leone 1972: 31–33).
199 Joannes Tzetzes, *Epistulae* 18 Τῷ μυστικῷ κυρῷ Νικηφόρῳ τῷ Σερβλίᾳ (ed. Leone 1972: 31–33).
200 Gregorius Nyssenus, *Tractatus secundus in Psalmorum inscriptiones* (ed. Migne 1863: XVI.604-XVI.605); *Digesta Iustiniani Augusti*, XXXXIII.10.5 (*De via publica et si quid in ea factum esse dicatur*) (ed. Mommsen 1870: 577.31–32).
201 Koukoules 1951: 249–336.
202 Ibn Battuta, *Rihla*, II.28 (eds Defrémery and Sanguinetti 1853–1858: 2.433).
203 Karpozilos 1989.

which was full of all kinds of waste disposed by the residents, is very indicative of the situation [204]:

ῥύπου τοίνυν τὴν λίμνην ἐνδελεχέστατα ἐνεπλήσαντο, νηχόμενοί τε καὶ πλυνοὺς ἐνταῦθα ποιού-μενοι καὶ ἀπορριπτοῦντες φορυτοὺς ἅπαντας...

So they kept filling the lake constantly with pollution, both swimming and washing clothes in it and throwing all manner of rubbish into it...

The Byzantine Empire was a heterogeneous and multinational mosaic shaped by historical events. As mentioned above about anthropo-geography, the frequent raids against the Empire forced the populations to gather around the nearest urban and semi-urban centers. The barbarian raids from the late fourth to the sixth century resulted either in the desolation of urban centers, or in the limitation of their size and area for the benefit of defense. The massive movements of peoples within the Empire never stopped due to the constant political and economic upheaval. They resulted in urbanism and forced urban planning abuses and informalities, which in turn, led to the deterioration of the health conditions of cities. To a large extent, the Byzantine cities were the next step in the evolution of ancient cities. This evolution, however, was far from uniform, but rather followed a varied course, imposed by local circumstances. In Byzantium, settlements appear to have been built according to the specific needs of the residents rather than according to concrete urban planning. It also appears that the old statutory building regulations gradually ceased to be applied, as no relevant officials existed to ensure their enforcement.[205]

Given that the plague epidemics in the Middle Ages were mainly of 'urban type', it is conceivable that cities and major trade centers would have been the primary target for outbreaks of serious infectious diseases. Based on the principles of modern epidemiology, disorderly urban planning coupled with overpopulation and the ensuing poor health conditions, have been one of the most significant risk factors involved in the outbreaks of infectious diseases.

Immunological Aspects of Plague

The process of historical and cultural evolution of the human species entails to a large extent the elements of dispersal of different population groups and their subsequent contacts, which occur either in a peaceful way or through conflicts. This evolution of interactions can also define the evolution of human ecology, which can in turn, and under specific conditions, change the standards of infectious diseases among human populations.

---

204 Procopius, *De aedificiis*,II.9.17 (ed. Haury 1913: 75.9 – 12).
205 Bouras 2012: 8 – 10.

The onset of the interaction between humans and microbes dates back to the time of the first socialization. Since the early age of agriculture and livestock, that is almost 10,000 years ago, we can identify three transition periods in the evolution of this relationship between humans and microbes. The period of the emergence of the first organized societies, which contributed to the spread of infections among the members of a community, can be considered as the first station. Thereafter, the extension of commercial networks promoted the exchange of infections between hitherto remote populations. Lastly, the Exploration Period that started in the fifteenth century and the consequent European expansion led to the exchange of even more deadly infectious diseases at an intercontinental level.[206] During that process, the *Homo sapiens* needed to survive against deadly microbes, and—in order to do so—the human organism had to develop a sophisticated defense system.

Life started on our planet 3.5 billion years ago, when unicellular organisms gradually evolved into more complex organisms. Approximately 600 million years ago, multicellular organisms began to form in combination with increased levels of atmospheric oxygen. This development was accompanied by a remarkable diversification of species, which has been termed as the 'evolutionary Big Bang'. The evolutionary lineage of vertebrates, which includes the human species, appeared 500 million years ago. During their evolutionary process, the superior organisms succeeded in developing, along with their other characteristics and properties, a unique immune system that is able to identify potentially fatal pathogens—including bacteria, viruses, fungi and parasites—and to initiate protective measures against them.[207]

Every organism is under constant threat by its surrounding environment. The aim of the immune system is to protect the host from the intrusion of malevolent organisms (microorganisms), as well as to prevent and hinder attacks from endogenous factors (like tumors and autoimmune diseases). The concept of immunity is divided into two basic types: innate and acquired (adaptive) (see Figure 1). The mammalian adaptive immune system developed approximately 500 million years ago in jawed vertebrates (gnathostomes).[208] The cells of these two types of immune systems communicate with each other during the immune response by secreting various immunomodulatory proteins (immunotransmitters), the so-called cytokines.

Innate immunity includes various physical barriers (skin and mucosal membranes that block the entry to most pathogens), as well as phagocytes (at the site of invasion), the complement system, and nonspecific inflammatory responses. The phagocytes are a type of white blood cells (leucocytes) that destroy foreign harmful particles through phagocytosis. Phagocytes include eosinophils, neutrophils, and macrophages. More specifically, macrophages have the ability to phagocytize foreign substances, while they also play a special role in the coexistence of innate and adap-

---

206 McNeil 1976: 103–104.
207 Cooper 2006: 815–822.
208 Laird 2000: 6924–6926; Flajnik and Pasquier 2004: 640–644; Cooper and Alder 2006: 815–822; Flajnik and Kasahara 2010: 47–59; Abbas 2012: 2–4.

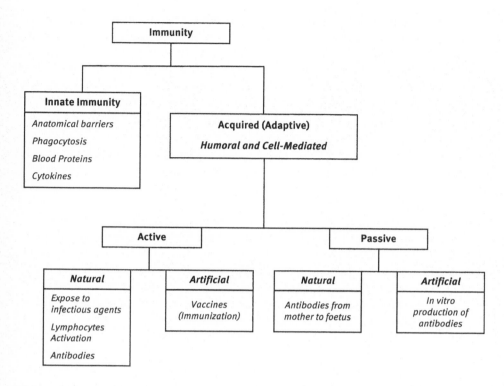

Figure 1. Types of immunity.

tive immunity. Intruders are also destroyed by a system known as *complement system*, which consists of a group of approximately 20 proteins circulating in the blood and extracellular fluids. One more element of innate immunity are the so-called *natural-killer cells* (NK-cells) that help phagocytes in destroying intracellular pathogens when their degradation is difficult, by producing special cytokines called *interferons.*[209] Together with another factor produced by lymphocytes, namely the tumor necrosis factor (TNF), interferons (IFNs) function as mediators of innate immunity. TNF-α and TNF-β proteins are referred to as TNF. The biological activity of TNF-α depends on the produced amount, and it aims to increase the secretion of special molecules and to activate the cells that destroy microbes.

The pathogens that manage to penetrate the physical barriers are recognized by the receptors of innate immunity. The set of mechanisms in this phase constitute the early induced immunity; they usually suffice to definitively cope with an infection. During this phase, certain products that are generated influence and define the evolution of the specialized immune response. At the same time, they give the organism

209 Abbas 2012: 350–352.

the necessary time to develop more specialized cells. Among the basic functions of innate immunity are nonspecific inflammatory responses at the point of invasion. Regardless of the type of the damaging factor and the site of damage, the response is standard. The purpose of cell concentration at the site of inflammation is the energetic uptake of foreign particles and microbes, and their subsequent intracellular digestion and destruction. The process starts with the entry of bacteria into the tissues, followed by a change in the vessel diameter and an increase in blood flow. Then a change in the vessel wall structure takes place, along with the output of plasma proteins, the migration of leucocytes at the site of the bacterial invasion, and the elimination of the damaging factor. Finally, a special signal of a cytokine informs the organism of the completion of the inflammatory reaction process.

If innate immunity fails to neutralize the exposure of the organism to foreign substances, other mechanisms are mobilized which are specific of some of these substances and significantly increase the defense capacity of the organism against possible future contacts with the same pathogens. These mechanisms, which constitute the adaptive immunity, involve the lymphocytes and the products they secrete, namely the antibodies, whereas the foreign substances that trigger the activation of this mechanism are called *antigens.* The form of adaptive immunity is called *active immunity* because the immunized individual plays an active role in the response against the antigen, in contrast to *artificial immunity* which is achieved by serotherapy and vaccination.[210]

Depending on the element involved in the immune response, adaptive immunity is distinguished into *humoral immunity* and *cell-mediated immunity.* Humoral immunity is characterized by blood molecules, the antibodies, the purpose of which is to identify and neutralize antigens. Antibodies are produced by a special subpopulation of lymphocytes, namely B-lymphocytes, which transform into cells that secrete antibodies upon contact with an antigen. Cell-mediated immunity involves another subpopulation of lymphocytes, namely T-lymphocytes, which usually interact with the phagocytes of innate immunity in order to destroy microbes.

The primary lymphoid organs of mammals are the thymus gland and bone marrow. All cells of the immune system originate from the bone marrow. The thymus gland differentiates T-lymphocytes. This is where the CD4 and CD8 lymphocytes are initially activated (they bear the name of the receptors on their membranes). The lymph nodes are strategically dispersed in the vascular system; they constitute the primary organs of the immune system that detect and prevent the spread of infections.[211]

All humoral and cell-mediated immune responses to foreign antigens are distinguished by the fundamental properties of specificity, which ensures that the immune response to a microbe will be targeted at each microbe specifically, as well as of di-

---

210 Abbas 2012: 2–4.
211 Shames 1995: 28–33; Abbas 2012: 2–13.

versity, which allows the immune system to respond to a wide variety of antigens. In addition to these properties, two other particular elements form the core of adaptive immunity. Firstly, the ability of cells to have an immunologic memory that increases their capacity to cope with recurrent infections from the same microbe. Secondly, the secondary immune response to a subsequent exposure to the same antigen is usually faster and often qualitatively different from the first immune response of the organism to this antigen. Immunologic memory occurs because any exposure to an antigen generates long-term memory cells which are specific to each antigen and able to survive for years after an infection.

Another basic element of adaptive immunity is clonal expansion, a property that helps increase the number of antigen-specific lymphocytes. In this way, the lymphocytes that are specific for an antigen proliferate significantly upon their exposure to this antigen.[212] It should be noted that upon the invasion of a microbe, the organism chooses the particular lymphocytes to be activated against the specific antigen. This fundamental capacity of the organism is called 'clonal selection hypothesis'. It was first proposed by the Danish immunologist Niels Kaj Jerne (1911–1994) in 1955 and was more explicitly conceptualized by the Australian immunologist Sir Frank Macfarlane Burnet (1899–1985), Nobel Prize "for the discovery of acquired immunological tolerance" (1960).

The adaptive immune response occurs in three phases: i) the recognition phase; ii) the activation phase; and iii) the effector phase. During the recognition phase, the antigens are connected to the special receptors on the surface of the B-lymphocytes. In the activation phase, the lymphocytes proliferate and are differentiated into antibody-producing cells. During the third phase, the activated lymphocytes neutralize the antigens. The lymphocytes that participate in this phase are called *effector cells*. Similar to B-lymphocytes, T-lymphocytes differentiate into specialized cells that activate phagocytes. In addition, the activated T-lymphocytes secrete protein-based hormones, which are called *cytokines* and enhance the action of phagocytes by instructing the inflammatory response.

The cytokines are another large and heterogeneous group of secreted proteins produced by many different cell types. The human genome includes approximately 180 genes that can encode proteins with the structural characteristics of cytokines. Many cytokines have been named on the basis of their biological activity, while some of them are called interleukins (IL).[213] During the activation phase, the macrophages undergo the effect of T-lymphocytes on cytokines, and simultaneously secrete a number of proteinaceous factors. The other cytokines that will, in turn, be excreted by the macrophages activate the necessary signal to cluster the inflammatory cells at the site of invasion. It is understood that the biological function of cytokines is an integral part of the defense system. More than a hundred structurally different and

---

212 Abbas 2012: 6–8.
213 Abbas 2012: 6–8.

genetically unrelated cytokines have been discovered. Despite their differences, they all have the following common characteristics: i) they are produced during the active phase; ii) their secretion is rapid but short-lasting and they tend to be self-limiting; iii) they have multiple different effects on the cells they are targeted at; iv) they bind to specific receptors on target cells; and v) they influence the composition and biological activity of other cytokines; as a result, some cytokines function as mediators of the activity of other cytokines.

The general pattern of the aforementioned elements is also activated in the immune response against *Y. pestis*. The immune response mechanism against *Y. pestis* is extremely complex and involves a combination of humoral and cellular factors. Individuals who survive the infection are considered to have acquired immunization against subsequent reinfections by *Y. pestis*, while the antibodies against specific antigens (*F1, LcrV, YopD*) can remain in the organism for more than ten years.[214] In the case of plague, innate immunity is inadequate against the disease. Post-mortem examination of victims of pneumonic plague identified plenty of bacteria, but little evidence of phagocytosis.[215] Studies in mice reveal a steady and progressive bacterial growth in pulmonary tissue and a spread of bacteria to other organs within 36 hours after infection. The lungs show remarkably little evidence of inflammation 24 hours after infection, while neutrophils increase only later on in the process. Thus, the phagocytes appear incapable of sufficiently controlling *Y. pestis* by themselves.[216]

Antibodies can undoubtedly better contribute and respond to the defense requirements against pneumonic plague. Early studies reached the conclusion that the natural defense mechanism against plague is primarily cellular, and that collaboration between the cells of the immune system was necessary for the effective destruction of bacteria.[217] Contemporary studies also show that the IFN-γ and TNF-α play an important role in the defense against plague. Thus, once the primary cytokines are released, they trigger a series of processes that offer an immediate protection against the infection and, if necessary, lead to adaptive immunity. Direct actions include the withdrawal of extracellular iron (IL-1 and TNF-α), increased endothelial-cell stickiness (IL-1 and TNF-α), mobilization and activation of professional phagocytes (IL-8), recruitment of neutrophils from marrow (IL-1), expression of reactive oxygen intermediates (IFN-γ και TNF-α), enhanced production of Fc receptors (IFN-γ), and initiation of granuloma formation (IL-1, IFN-γ και TNF-α).[218]

It is generally accepted that the principles and basic mechanisms of the immune response of the human organism have remained unchanged for thousands of years and that the immune response of medieval populations was essentially the same as

---

214 Yujing 2016.
215 Lien-Teh 1926; Smiley 2008.
216 Lathem 2005; Lathem 2007; Bubeck 2007.
217 Meyer 1950; Jawetz 1944a: 1–14, and 1944b: 15–30.
218 Brubaker 2003.

today. But what are the factors that determined that some died and other survived during an epidemic? A possible answer to this question is the individualized immunological profile of each person. Given that cytokines like IFN-γ or IL-10 are targeted by *Y. pestis* upon its invasion, the various levels of their production in each individual can ultimately prove to be a critical factor for an active immune response.[219] Studies based on databases which can show—but only to a certain extent—the general trends in the immunological profile of human populations in various geographic regions, have revealed interesting facts regarding the levels of IFN-γ and IL-10 production. Especially for IL-10, it is worth noting that the regulation of the inflammatory response is one among its main biological functions. Studies show that the antigen *LcrV* of *Y. pestis* aims to inhibit the action of this cytokine. The deregulation of IL-10 results in a false signal that the inflammatory response has been completed.[220] As a consequence, the response of the innate immunity at the site of invasion remains incomplete, allowing *Y. pestis* to continue its devastating work. The analysis of databases has revealed that European population in different regions of the same country (namely in England and Italy) were specified; their baseline production of IFN-γ and IL-10 was different.[221] Although these findings are certainly not automatically transferable to medieval populations, they may imply the existence of a genetic variation of cytokines at that time.

The affirmation of Byzantine writers according to which the "living did not suffice to bury the deceased" may have depicted the real picture of an epidemic. Nevertheless, this situation could be explained in the context of a massive exit of the survivors from the cities. The lack of reliable demographic data on Byzantine cities raises questions as to the real case of fatality and mortality rates of the disease. This essentially points to the next question regarding the phenomenon of a different case fatality rates, or overall different mortality rates among various cities. The phenomenon of different mortality rates has been better studied in the case of Western European epidemics during the *Black Death*. Despite the lack of Byzantine sources we can conjecture that this might have been the case in Byzantium, too. Further examination of this phenomenon raises another question with regard to the relationship between the epidemic outbreaks of plague and the possible degree of immunity that medieval populations developed against the disease.

At this point, another parameter of immunological and epidemiological relevance must be considered. Among the basic assumptions of medico-historical studies is the equal risk of members of a population group when exposed to the same infectious agent. In fact, an epidemic never affects an entire population, but, even if we assume that all members of the population are equally in danger of being exposed to a disease, some of them will be prepared for a different and more effective

219 Crespo 2014.
220 Nedialkov 1997; Sing 2002a and 2002b.
221 Turner 1997; Couper 2008.

immune response, while others will not react successfully to the invasion of the infectious agent for a number of reasons.

The immune response of an individual is influenced by numerous parameters, most notably the existence of a pathological condition (e. g. malignancy) or a co-infection that results in the suppression of the capacity of the immune system.[222] Given the history of the particular disease and its common evolution along with humanity —until its eventual treatment in the twentieth century—, the case of tuberculosis will be called upon. It dates back to the first human societies. The microbe of the disease, *Mycobacterium tuberculosis*, has been identified in paleopathological findings of human fossils nearly 3,500 years old.[223] The finding of mycobacteria in human fossils essentially covers the entire chronological spectrum of human history, including the medieval period. Given the diachronicity of *M. tuberculosis*, it is assumed that a significant part of the European population had been infected and died of tuberculosis during the Middle Ages. The studies on the impact of *M. tuberculosis* on the immune response of patients against other infectious diseases confirm its aggravating role in reduced immune response.[224]

Another possible cause that leads to reduced immune response in modern humans is alcohol consumption, especially alcoholism. The consumption of all kinds of alcoholic beverages has been a timeless component of human history and, although it is extremely difficult to compare alcohol consumption in modern societies with that of the medieval period, many researchers suggest that alcohol consumption was an integral component of medieval societies as well.[225] Both acute and chronic alcohol consumption can affect immune responses by reducing antimicrobial defense against bacteria and viruses. Chronic abuse of alcohol has been associated with immunosuppression, increased incidence of infections (especially bacterial pneumonia), as well as increased morbidity and mortality.[226] Alcohol abuse is also related to increased occurrence of various serious infections, such as tuberculosis and chronic hepatitis C infection.[227] Last but not least, we must not forget the existence of people with primary immunodeficiencies, which constitute a heterogeneous set of diseases that, depending on the patient's age, are characterized by immunological deficits of various degrees. According to the WHO classification, these deficiencies may be related to B-lymphocytes, T-lymphocytes, phagocytes, or the complement system.[228] Apparently, the nosological spectrum of the Middle Ages did not only include plague, but various other infectious and non-infectious diseases as well. As

---

222 Djurdjevic 2009.
223 Donoghue 2008.
224 Nissapatorn 2006; Shaddock 2014; Moreira-Texeira 2017.
225 *Official Publications Received* 1933; Bales 1946; Warner 1992; Adamson 2004: 48–53; Roberts 2007.
226 Cook 1998; Szabo 2009; Molina 2010.
227 Wiley 1998; Prakash 2002; Zhang 2008.
228 Al-Herz 2012. Also World Health Organization 2015.

with modern humans, the existence of a pathological condition in human populations of the past could lead to reduced immune response, making a person susceptible to the (new) pathogens that entered the body.

The existence or absence of an infectious disease in an individual who has been exposed to an infectious agent depends on the individual's susceptibility or resistance to the specific infectious disease. However, immunity is a relative concept that can be reduced significantly when a long time has elapsed since infection or when the individual's immune system has been attenuated.

Individual immunity is a powerful force that affects both a person's health and the evolution of pathogens. If we generalize individual immunity and adopt it for a larger set of people, then the immunity of this social group is defined as *herd immunity*. Due to the upward dynamics of infectious diseases, an individual's infection simultaneously increases the risk of infection for other individuals of the population. On the contrary, if the share of immune population (as a result of vaccination or of a previous infection) exceeds the share of susceptible individuals, then the impact of the pathogen decreases. This decrease occurs because the microbe invades individuals who had been infected (or vaccinated) in the past and—as a result—are immune and shall not transmit the disease. Thus, herd immunity changes the dynamics of transmission of pathogens, and epidemics may be slowed down.[229]

The specification of the 'critical rate' of immune individuals is particularly important, that is, the rate of herd immunity that is sufficient to protect a population. This rate depends on the infectivity of the infectious agent, the duration of the infectious period, the number of sources of infection, the frequency of introduction of the infectious agent in the population, the density of the population, and the health consciousness of this population. Despite the ability to acquire individual and consequently herd immunity, the level of immunity decreases over the years as the composition of the population changes, as a result of the addition of new individuals due to births, or the movement of populations.

The significance of herd immunity in protecting a population against infectious diseases is apparent in the cases of epidemics that occur when an infectious agent invades an immunologically 'virgin' population, i) which has never come in contact with the infectious agent in the past, or ii) when the contact occurred in the distant past and the immunity acquired by the population has faded out, or iii) when the infectious agent has undergone a mutation. Instances of invasion of immunologically virgin populations are characterized by extremely high rates of morbidity and fatality. A typical example of such devastating epidemics is the virus of smallpox. For instance, the native populations of the African and American continents were decimated during the era of European explorations and colonization.[230] Another typical example was the smallpox epidemic that broke out in Iceland in 1707, which super-

---

229 Metcalf 2015.
230 McCaa 1995; Hays 2005: 131–134; Schneider 2009; Taylor 2014: 146–153.

seded another epidemic that had broken out 35 years earlier. The epidemic decimated the individuals born after the previous epidemic of 1672 who were not immune, since they had never come in contact with the virus.

Similarly, the epidemics of Byzantium constitute a typical epidemiological model that comprises all the aforementioned elements. The anthropogeography of the Empire reveals the presence of a dynamic population system during the Byzantine era, which was characterized by constant inflows of new individuals. Population movements and refugee flows by millions are a common phenomenon nowadays. At the health level, however, proper epidemiological surveillance coupled with vaccination programs in host countries result in new immune individuals. Thus, the vaccination of these populations allows them to acquire artificial active immunity, to retain a high level of herd immunity, and to strengthen their immune barrier. Although it is impossible to estimate the level of herd immunity in Byzantium, we may simulate the natural evolution of herd immunity at a time without vaccines and thus without artificial active immunity and the creation of new immune individuals.

The first affected Byzantine populations should have been almost completely decimated, but, based on the epidemiological and immunological standards, it is obvious that a group of individuals always survived (excluding those individuals who survived simply because they left their places of residence) and acquired natural active immunity that defended them against future contact with the disease (the exact duration of immunity remains unknown) and turned them into a barrier to the spread of the disease (at least within the limits of their city/region). Nevertheless, population movements, due to either invasions or decisions to relocate them to the affected areas in order to fill the demographic gaps, essentially created a new set of 'virgin' populations. We may assume that the share of people who survived was perhaps so small that immune individuals were not sufficient to protect the susceptible newcomers in subsequent disease outbreaks.

Byzantine history is full of such movements that introduced new populations to places already affected by the disease. During the fourth century, the eastern part of the Empire had already been struck by invasions and major migrations of peoples, which resulted in a demographic decline in the provinces along the southern bank of the Danube. This gap was gradually compensated for by the establishment of Visigoths in the north of the province of Thrace, and the Ostrogoths in Pannonia. In addition to the major movements and settlement of new tribes in the Empire, internal migration was intense at the time of plague. The epidemic waves of the sixth century led to a decline in the population, although it is difficult to estimate its exact rate. But the easy settlement of Slavic populations and the raids of the Avars that followed the epidemics raise suspicions on the high fatality rates and devastation caused by the disease.

An important demographic event that took place during the seventh century was the descent and settlement of Slavs in the Balkan Peninsula. Their history is connected with that of the Avars. The Slavs seceded from the domination of the Avars to join the Byzantine army and gradually expanded their settlements in the Balkans, in an area that still remains to be exactly identified, particularly its southernmost limits. At the time of the Slavic settlements, the Vlachs moved to the Empire.

During the eighth century, the epidemics had apparently decimated the population of Constantinople, as made clear by the resettlement of populations from Central Greece, the Peloponnese and the Aegean islands to the capital to fill the demographic gap in the city. At the same time, the conquest of the Byzantine Middle East by the Arabs forced thousands of people to abandon the area and to resettle in other provinces of the Empire. The loss of Egypt and Syria was coupled with that of significant and crowded commercial centers like Alexandria and Antioch. The void of the deserted borders left behind by the long-lasting wars between the Byzantines and Arabs was essentially filled by the resettlement of refugees. Moreover, as a result of the loss of the Middle East, residents of entire cities and various tribes of people fled to Asia Minor.

In addition to displacements due to wars, state-planned movements of populations to areas that needed demographic support should be mentioned. The selection of the populations to be moved was based on political, economic, or religious criteria. The imperial decisions of Constans II (imp. 641–668) and Constantine V (imp. 741–775) to move parts of the Slavic populations to Asia Minor are typical examples of this. The decisions of Constantine V and, further on, of Leo IV (imp. 775–780) fall within the context of such organized population movements, which resettled several thousands of Christians from Asia Minor to Thrace. Lastly, we should not forget the cases of 'ecological refugees' who abandoned their homelands mainly because of long periods of drought.[231]

These and many more movements took place between the epidemic waves that struck the Empire. The areas affected by the disease may well be likened to volcanoes. As in major volcanic eruptions—the incidence of which is unknown—, it is not known when the disease would be transmitted to humans and turn into an epidemic in enzootic regions. Therefore, when the conditions were ideal for the reemergence of an epidemic, the presence of new susceptible individuals at that time and in that place simply translated into a new carnage. We may well assume that the movement and settlement of new 'virgin' populations in areas with a burdened epidemiological past were an additional factor that could have triggered the swift spread of plague.

Climate as a Trigger Factor

As already mentioned in the introduction of this chapter, the environment is a key factor of the epidemiologic triad. The environmental factors often influence and determine the spread of infectious diseases and epidemics. Water resources, crops, and

---

231 Koder 1984: 135–154.

climate constitute three important environmental parameters in this relationship. Nowadays, climate change is estimated to cause approximately 150,000 deaths annually, including deaths due to extreme weather conditions. Changes in temperature and the rainfall rate can also influence the patterns of transmission of numerous infectious diseases associated with waterborne infections and infections that spread via transmitters. The distribution and size of the vector population of microorganisms can also be greatly influenced by the local climate and its changes.[232] The relationship between vector-borne diseases and climate has been known since long ago, and its study helps to predict the behavior of epidemic diseases.

Weather and climate are two terms that actually represent two aspects of the same spectrum. Weather is defined as the state of the atmosphere at a given point in time, which covers the evolution and life cycle of the specific atmospheric disturbances. Climate is the average weather condition of the atmosphere and the surface of the earth and seas, within a specific area and a given time span.[233]

Climate is usually described by the comprehensive statistics of a set of variables of the atmosphere and the surface of the earth, such as: temperature, wind, humidity, soil moisture, sea surface temperature, and sea ice concentration and thickness. Based on the international standards, a minimum of 30 years is considered as an adequate period for climate analyses, although periods between 10 and 15 years can also provide sufficient data. Climate classification is an extremely complicated task that can be achieved on the basis of vegetation and plant growth (Koppen classification), energy and water balance (Thornthwaite classification), general air circulation (Flohn and Strahler classification), and the discomfort index, that is, the thermal equilibrium between humans and the environment (Steadman classification).

Microclimate is an equally common term used to define the climate of a smaller or restricted area within a broader region, which results from the influence of the terrain, the soil surface, and other factors like the difference between air and soil temperature, or the humidity or wind intensity. Notably, the microclimate of slopes, valleys and hills is characterized by the elements of the terrain of an area. More specifically, the microclimate of slopes depends on their exposure and inclination to the incident solar radiation. Moreover, the microclimate of meadows and spaces surrounded by trees or forests, as well as of coastal regions or lakes, is the result of the different heating of the underlying surface by radiation. Instead, the microclimate of a city varies according to urban construction and development of the residential web.[234]

Regional climatic changes with less direct mechanisms could affect the transmission of many infectious diseases and also the rate of food productivity (especially of cereal and grain). As far as vector-borne infections are concerned, the distribution

---

232 McMichael 2003: 543.
233 McMichael 2003: 543.
234 Orlanski 1975; Salinger 2005; Tsiros 2009.

and abundance of vectors and hosts are influenced by various factors, be they natural (temperature, rainfall, humidity, surface water and winds) or biotic (vegetation, host species, predators, competitors, parasites and human interventions).[235] Both the infectious agents and their carriers lack thermostatic mechanisms and their survival is thus, directly influenced by the local climate. Therefore, there is a limited range of climatic conditions within which any kind of infectious disease or carrier can survive and reproduce. The pathogens transmitted through carriers spend part of their life cycle within cold-blooded arthropods that are subject to many environmental factors. Changes in weather and climate may influence the transmission of vector-borne diseases and include temperature, rainfall, wind, extreme wear or drought, and sea level rise. The pathogens transmitted via rodents can also be indirectly influenced by ecological factors. The impact of temperature on rodent populations is less noticeable since rodents are homeothermic, and do not respond immediately to temperature fluctuations of the environment. In temperate areas, however, low temperatures in winter can adversely affect rodent populations because of the sparse availability of food.[236]

Apparently, the balances are quite subtle and critical but—most importantly for the ecological culture of modern humans—the results of environmental changes can have an immediate impact or emerge in the long term, and affect future generations. Although they may appear as unlikely scenarios, environmental changes create a fertile ground for the future outbreak and spread of a disease.

As expected, climate affects all three components of the plague system (bacteria, vectors, and hosts) and may be a possible factor to explain the variability of plague dynamics which can transform the disease from a small and regional epidemic to a large-scale one.[237] Since the early twentieth century, climate has been characterized as a basic factor in the alternation of recession and outbreak of the disease. Already in 1928, the Major General Sir Leonard Rogers (1868–1962) proposed to consider seasonal fluctuations in temperature and humidity to understand the pattern of outbreaks of plague in India.[238] A few decades later, it was found that plague foci in several African countries were also heat-dependent.[239] Subsequent studies also showed an increased incidence of plague in Vietnam during the hot and dry seasons that follow the periods of high seasonal rainfall.[240] In fact, these studies also revealed a unique relationship between plague and rainfall, according to which the risk of plague increased during dry periods and when rainfall was less than 10 mm.[241] The epidemiological cluster of plague is the result of complex interactions between the life

---

235 Patz 1996; Kovats 2001.
236 Korslund 2006.
237 Ben Ari 2011; Büntgen 2012.
238 Rogers 1928.
239 Davis 1953.
240 Cavanaugh 1968 and 1972.
241 Stenseth 2006; Pham 2009.

cycle of its vectors, and their dynamics and geographical distribution, which are influenced by climate variables. Indeed, temperature, rainfall, and relative humidity have a direct impact on the growth and survival of fleas and their populations, as well as on their behavior and reproduction.[242] For instance, the rate of development of *Xenopsylla cheopis* from an egg to an adult flea is regulated by temperature and relative humidity.[243]

In Peru, fluctuations in climate stemming from the El Niño phenomenon were related to the appearance of plague in 1999, just like in Northern Colorado with the epizootic of wild dogs in the area.[244] In most natural foci of the southwestern part of the Western United States the outbreaks of plague were also related to the increased rate of rainfall during winter and spring.[245] In New Mexico, plague cases among humans were more frequent after winter and spring periods with extremely high average rainfall. The "trophic cascade hypothesis" was formulated to explain these findings. According to it, increase in rainfall often leads to an increase in the demand for food resources that results in an increase in rodent populations. In some cases, the trophic cascade hypothesis seems to explain the increase in plague epizootic, such as the black-tailed prairie dog *Cynomis ludovicianus* host in the Southwestern United States.[246]

In the case of Asia, it appears that the cases of plague among both humans and rodents have been associated with climate variability. Studies have shown that the epizootic of large gerbils (*Rhombomys opimus*) in Central Asia was higher during the years with wet summers and warm springs. An increase of 1 °C during spring leads to an increase of 50% in the prevalence of plague among gerbils.[247] In China, the cases of plague among humans have also been linked to the climate changes associated with the southern oscillation index and the sea surface temperature of the tropic Pacific east equator.[248] As far as China is specifically concerned, model analysis shows that the disease is directly dependent upon the dry or wet conditions of various parts of the country. Based on the historically recorded epidemics, the prevalence of plague in North China increased along with humidity, whereas, in South China, plage subsided when humidity increased. These conflicting facts are probably associated with the different climate of North and South China, and the different families of rodents that live in these two regions.[249] In any case, the relationship between plague among humans and climate change has not yet been fully understood, whereas few studies exist that explore both tropical and subtropical zones.

---

242 Krasnov 2001a, 2001b, and 2002.
243 Bacot 1924; Gage 2008.
244 Stapp 2004; Ben Ari 2008.
245 Parmenter 1999; Enscore 2002; Ben Ari 2010.
246 Xia 1982; Singleton 1989; Brown 2002; Davis 2005.
247 Stenseth 2006; Kausrud 2007; Snäll 2008.
248 Zhang 2007.
249 Xua 2011.

Frequent attempts have been made to interpret all the aforementioned factors in order to explain the interactions of the past. The indirect information on climate derived from the study of tree rings to specify the exact year they were formed (dendrochronology), of plant pollens in archaeological layers (palynology), of sediments and deposits of rivers and lakes (sedimentology) or glaciers (glaciology), as well as of archaeological evidence (buildings, settlements, road networks, for instance), contribute to the broad multidisciplinary database of proxy data on climate history. Researchers are divided about the various hypotheses of climate evolution during the first and second millennia. Many support the hypothesis of gradual drying of the climate and pulsatory climatic changes between warm and cold periods in Eurasia. On the other hand, numerous researchers are in favor of the model of relative climate stability with short- or medium-term alternations between dry and wet periods. In addition, an alternative theory suggests the change towards a more humid climate in the Mediterranean basin occurred during the fourth century A.D.[250] According to modern data, the Mediterranean climate is characterized by very warm and wet summers, relatively mild winters, and a varying rate of rainfall. This is a distinctive characteristic of the Eastern Mediterranean, which is generally associated with lower rates of rainfall compared to the rest of the Mediterranean. The interior of the Balkan Peninsula is characterized by a mild continental climate with harsh winters accompanied by heavy frosts and increased rainfall. In Upper Mesopotamia, which is connected with Asia Minor to the southeast, the northern and southern parts have different climates. While rainfall in the north of Mesopotamia is a common phenomenon, it is scarce in the south. Winters are cold in the north and mild in the south, and summers are rather warm and dry with slight differences between the north and the south.[251]

Very little data exist for the Early Byzantine period (fourth and fifth centuries) to possibly simulate climate changes and their effects. As for the sixth century, the scarce available data could be used to simulate its climate up to a certain extent. It appears that the drought of the first decades of the sixth century was followed by years of increased rainfall.[252]

The evidence stemming from the study of Byzantine sources dating from the seventh century suggests the existence of drought. Reports of periods of drought are common during the period 500 – 750 A.D., and are mainly focused on Egypt, Palestine, Syria, and Mesopotamia. Periods of drought were also reported in Asia Minor and Thrace (Constantinople) during the sixth century A.D. Generally speaking, the period between the fourth and the seventh century is characterized by generalized drought in the Eastern Mediterranean and the Middle East. According to the estimates of paleoclimatologists, the evolution of climate in the Eastern Mediterranean

---

250 Chappell 1971.
251 Koder 1984: 135 – 154.
252 Lamb 1977: 262.

basin appears to have been coherent. Climate seems to have been similar to modern climate, maybe warmer and drier until the Early Byzantine period. It became a little more humid in the following centuries, but continued to alternate between wet and dry periods. The numerous irrigation works that were carried out in the plains of Mesopotamia for the collection of water seem to validate the hypothesis of more humid climate conditions during the Early Byzantine and the Early Islamic periods. Similar theories were proposed on the basis of the number of rainwater harvesting buildings in the Negev desert and northeastern Sinai in the Palestinian region during the Byzantine rule. The first indications stemming from the few dendroclimatological data for this area point towards periods of fluctuating drought. Moreover, pollen samples from Syrian sediments also suggest a dry climate in the area.[253]

The increasing number of reports in the following centuries reveals that the period between the eighth and the thirteenth century was characterized by a colder and more humid climate in the Eastern Mediterranean and the Middle East. More reports are available for Constantinople, Asia Minor, Armenia, and Mesopotamia. Arabic sources also mention dry periods in Egypt, Syria, and Palestine. The alpine glaciers began to expand ca. 1300, marking the beginning of the so-called 'Little Ice Age' which seems to have affected the wider Mediterranean region, since the Byzantine sources speak of a cooling of the climate. References to cooling in the areas of Constantinople, Asia Minor, Thrace and Macedonia increased in the period 1300 – 1400 A.D.[254]

All these climate changes must have affected—to a certain extent—the evolution of infectious diseases in the wider region of the Eastern Mediterranean. With regard to the dynamics of plague, it could be argued that the climatic conditions of the sixth century perhaps proved ideal for the establishment of the microbe in the Mediterranean, whereas the changes that occurred in the following centuries contributed to its preservation in the region. The following cases seem indicative: the periods of drought and decreased rainfall in Palestine between 516 – 520 and 523 – 538 followed by rainy and wet years in the 540s, the dry and hot years of 536 – 541 in Mesopotamia, and the drought and plague that affected Constantinople and Asia Minor between 554 – 562.[255] Modern studies indicate that the climate can influence plague both directly and indirectly, and involve the infectious agent, the transmitter, and the host. The studies of the twenty-first century show that the population and behavior of rodents is influenced by climate changes. Thus, how could anyone deny that the same pattern determined the ethology of rodents in the past centuries? The accounts of intense weather phenomena and the subsequent plagues that affected both humans and animals show that one of the three aspects comprising the epidemiologic

---

253 Telelis 2004: 850 – 864.
254 Telelis 2004: 850 – 864.
255 Telelis 2004: 850 – 864.

triad of plague, namely the environment linking the pathogen to the host, had already been disturbed before the appearance of the disease in the Mediterranean.

Chapter 4
Plague-resembling Epidemics from Antiquity to Byzantium

When ancient peoples referred to an epidemic outbreak, the terminology that they used did not describe the disease itself, but rather its consequences, namely the number of deaths and the catastrophe it caused. For instance, an epidemic outbreak is referred to as *Shechin* in Hebrew, as *Ta-ūn* in Arabic, as *Mari Mahamari* in Hindi, as *Marga Margi* in Persian, and as λοιμός (*loimos*) in Greek.[256] Regardless of the exact names, however, such terminology always represented the fear of a widespread and deadly epidemic. In the Byzantine world, another term appeared besides the aforementioned *loimos:* θανατικόν (*thanatikon*) which translates into 'deadly', 'widespread death'. Over the centuries, the meanings of *loimos* and *thanatikon* essentially became indistinguishable from the notion of plague. Although the Anglo-Saxon literary word 'plague' is widely used in translations of ancient texts, it is hard to establish its relationship to the disease caused by *Y. pestis*, since it does not shed light on the exact nature of the infectious disease that affected a city or a region. In fact, the word 'plague' may point to any infectious disease that spreads rapidly and has a high mortality rate.

As a medical term and irrespective of whether it is used generically or in order to describe the disease caused by *Y. pestis*, the word 'plague' stems from the Greek word πληγή (*plêgê*), which translates into a 'blow'. Similarly, in Latin the same word was used to describe a 'stroke' or a 'wound'.[257] On the other hand, the Latin word 'pestis' denotes a disease, a ruin or destruction caused by a plague or a fire, for example. Medieval chroniclers typically use termins like *mortalitas, pestilentia, epidemia*, but the words choice seems to have little significance. At the same time the epidemic event was characterized as "universal plague" or "universal pestilence".[258]

The confusion is even more pronounced when the reader realizes that the Ancient Greeks used the word λοιμός (*loimos*) to describe any disease that resulted in a large number of victims. At the same time, however, another word exists in the Greek language to describe a destructive disease, πανώλη (*panole*). The word derives from the ancient Greek words πᾶς (*pas*, meaning *all*) and ὄλλυμι (*ollumi* meaning *to destroy*), in other words, it tells us that *everybody is destroyed* (ruinous, deathly, completely destroyed).[259] Over the ages, the words λοιμός and πανώλη of the Ancient Greeks became indistinguishable from the Byzantine θανατικόν (*thanatikon*), where-

---

256 Khan 2004.
257 Taylor 2017: 29.
258 Valpy 1828; De Vaan 2002; Garmichael 2008:17–19.
259 Montanari 2015: sub verbo.

https://doi.org/10.1515/9783110613636-007

as all three words were used in Greek texts to describe the disease defined as plague by the Anglosaxons, overshadowing the reports of other infectious diseases.

Various reports dating from the Antiquity and Early Middle Ages refer to epidemics, the etiological factors of which have remained medico-historical mysteries until today. Nevertheless, the selective study of symptoms, whenever they are mentioned, has led numerous contemporary researchers to the conclusion that these were cases of plague epidemics that occurred in Byzantium before the sixth century.

The *Plague of the Achaeans* (12th c. B.C.)

The infectious disease referred to as the *Plague of the Achaeans* in Rhapsody A (I) of the *Iliad* could be associated with plague only semantically.[260] During the siege of Troy, the god Apollo, enraged by the king Agamemnon who had insulted Chryses, the god's priest, sends with his bow and arrows an epidemic on the Greek army. And in fact, among the various theonyms given to the god Apollo by the Greeks was Σμινθεύς (*Smintheus*), that is, *the one who brings mice*. Some scholars believe that this theonym stems from a distant memory of a plague epidemic.[261]

The *Plague of the Philistines* (11th c. B.C.)

Another reference to a possible outbreak of plague can be found in the Old Testament and is the *Plague of the Philistines*.[262] According to many researchers, this is the first reference of the existence of plague in the Ancient World. Based on the Old Testament book *First Kings (First Samuel)* (4–6), prior to the battle against the Philistines, the Israelites transferred the Ark of the Covenant from the city of Shiloh to their camp in Eben-Ezer. Upon the defeat of the Israelites in the battle of Eben-Ezer (mid-eleventh century B.C.) and the subsequent looting of their camp by the Philistines, the Holy Ark was captured. The narrative has it that the wrath and vengeance of the God of Israel was great on those who dared to desecrate the Ark. During the seven-month period that the Philistines possessed the Ark, a terrible epidemic broke out in their territories. They then began to transfer the Ark from town to town, in a desperate attempt to stop the epidemic. Nevertheless, wherever the Ark arrived, the divine wrath ensued. Gaza, Gath, Ashkelon, Ekron, and mainly Ashdod were the five cities that mourned thousands who died as a result of the desecration of the Holy Ark. The Philistine priests eventually speculated that the return of the Ark to the Israelites would atone for the desecration and save their people from death. They thus

---

260 Murray 1954: 2–4.
261 Berheim 1978.
262 *Vetus Testamentum*, *Kings I/Samuel* I.4–6 (ed. Rahlfs 1965: 510–513).

returned the Ark to the Israelites, and the disease disappeared from the Philistine territories. The Biblical text reads as follows in its English translation:

> But when they transferred it (the Ark) there, the Lord made the residents of the city feel His great strength upon them. They were all struck by swellings in the groin, both the young and the elderly.

The 'clinical' information conveyed by the text can by no means be isolated to support the diagnosis of plague, although many researchers associate the groin swelling with the inguinal lymph nodes.[263] The text offers another hint that supports the view that this was the first recorded case of plague. The priests and sorcerers urged the Philistines to offer gifts to the Israelites upon the return of the Ark, in order to make amends for the sacrilege. The gifts offered were five pairs of golden artifacts, one from each Philistine city, and corresponded to one golden model of a swelling and one golden model of a mouse.[264]

It is conceivable that the correlation of the golden offerings with the given role of rodents in the epidemic cycle of the disease has led numerous researchers to identify the *Plague of the Philistines* with plague.[265] At the same time, the theory of tularemia has been proposed, as its epidemic cycle involves rodents too, as well as the theory of bacterial or parasitic dysentery.[266]

The *Plague of Athens* (430 B.C.)

The most complicated event, however, which has been widely regarded as an epidemic of plague, is the *Plague of Athens* or *Plague of Thucydides*. As depicted by the Greek historian Thucydides (ca. 460–ca. 400 B.C.) in his narrative of the Peloponnesian War, the *Plague of Athens* is possibly the greatest enigma in the history of medicine.[267] According to this description, the plague suddenly appeared at the port of Piraeus during the second year (430 B.C.) of the war between Sparta and Athens. The plague is said to have originated in Ethiopia and to have then spread to Egypt, Libya, and Persia. From Egypt it reached the port of Piraeus, and it quickly spread to the city of Athens. At first, the Athenians believed that the Spartans had poisoned their wells, but they soon realized that they were dealing with a deadly disease.

According to Thucydides, the patients suffered of high fever in the beginning, with headache, eye redness, sternutation, and cough, while the oral cavity and the

---

263 *Vetus Testamentum, Kings I/Samuel* I.5 (ed. Rahlfs 1965: 510–513).
264 *Vetus Testamentum, Kings I/Samuel* I.6 (ed. Rahlfs 1965: 510–513).
265 Gwilt 1986; Trevisanato 2007.
266 Haeser 1882: 4–24; Mollaret 1969; Dirckx 1985.
267 Thucydides, *Historiae,* II.48–54 (ed. Hude 1901: 171–177).

pharynx were bloody. The body was very red and characterized by a generalized pustular rash. The last stage of the disease was characterized by a profuse diarrhea, and the patients passed away after seven or nine days from the onset of the symptoms. Thucydides also describes a situation that could be identified with delirium. Equally interesting is the report on the decimation by the plague of the Athenian expeditionary force in Potidea, in July 430 BC, on which various mathematical models have been based in an attempt to define the etiological factor of the plague.

The *Plague of Thucydides* has been extensively studied and many hypotheses have been proposed concerning its nature. Unfortunately, given the lack of archaeological findings that could directly point to microbiological data, these theories are based solely on the description by Thucydides and the isolation of specific symptoms that lead to a number of possible disease entities. For instance, plague, smallpox, anthrax, typhus, ergotism, tularemia, Ebola, and flu have been proposed.[268] But, the sudden discovery of a mass grave containing at least 150 skeletons in the Kerameikos ancient cemetery of Athens rekindled hopes for the solution of the ancient medical conundrum.[269] The application of molecular analysis has yielded interesting, yet questionable results. The implementation of PCR on genomic parts of six pathogens, namely of plague (*Y. pestis*), typhus (*Rickettsia prowazekii*), anthrax (*Bacillus anthracis*), tuberculosis (*M. tuberculosis*), cowpox (cowpox virus), and cat-scratch diseases (*Bartonella henselae*), did not yield any product in the reaction with the DNA samples obtained from three ancient teeth. The pulp of these teeth was found to contain DNA sequences of *Salmonella enterica* serovar Typhi, and thus typhoid fever was considered as the most probable cause of the *Plague of Athens*.[270]

Although this diagnosis seems plausible, we cannot exclude the scenario that the Athens sequence is probably related to the modern, free-living soil bacterium. In fact, the debate focused on the genetic proximity of the Athens sequence to the *Salmonella* species. The simple phylogenetic analysis shows that the ancient sequence does not relate to *S. enterica* and *Salmonella typhimorium* or various other species of *Salmonella*.[271] In simple terms, this story reveals the authentication of results as one of the most important and recurring problems in the research of ancient DNA.[272] The phylogenetic verification of results is still ignored, despite the strict laboratory guidelines and technical protocols.

Prior to the adoption of molecular techniques in medicine, an interesting epidemiological approach to the possible disease is worth mentioning. According to Thucydides, in a demonstration of high strategy, the Athenians led a campaign against Potidaea, a prominent colony of the Spartans in Chalkidiki (northern Greece), in

---

268 Morens 1992 and 1994.
269 Sherman 2017: 53–56.
270 Papagrigorakis 2006; Chunha 2008.
271 Shapiro 2006.
272 Cooper 2000.

order to force the besiegers to withdraw part of their army from Athens. Among the 4,000 men comprising the Athenian expeditionary force, however, some had already been infected and the first symptoms began to appear during the siege of Potidaea. As a result, 1,050 Athenian soldiers died within 40 days. The epidemiological question that arises is what disease could have infected those men in that 40-day period.

The approach was based on the correlation between humans and the disease as defined by a mathematical relationship. In this particular case, the idea originated from the classical Bernoulli trial, which is a random experiment with two possible outcomes that are mutually exclusive, 'success' and 'failure'. Let the probability of success be represented as p and the probability of failure as q, then

$$p + q = 1.[273]$$

When this basic concept is used in the case of a human disease, then a healthy individual who comes into contact with an individual suffering from a serious infectious disease has a 50% chance (p = success) of becoming infected and 50% chance of not becoming infected (q = failure). But when a healthy individual contacts more than one infected individual—and based on the multiplication rule of probability—then, the probability of becoming infected rises at the expense of remaining healthy. Thus, an attempt was made to determine the incidence of new outbreaks at any given phase of the epidemic according to the Reed-Frost model, which constitutes one of the principles of epidemiology:

$$C_{t+1} = S_t[P_{t+1}]$$

t represents a time period

C_t represents the number of infected individuals at time t

S_t represents the number of susceptible individuals at time t

P_{t+1} represents the probability that any two individuals randomly selected will come into effective contact (that is, contact sufficient for an active case to infect a susceptible individual)

also, $P_{t+1} = (1 - q^C_t)$

q represents the probability of inadequate contact between two individuals.

It should be noted that epidemiological models rest on certain assumptions, the most common of which are:
a) the disease is transmitted by contact between an infected individual and a susceptible individual;
b) infection occurs instantaneously upon contact;

273 Mathur 2010: 84; Trivedi 2016: 47–64.

c) all susceptible individuals are equally susceptible and all infected individuals are equally infected;

d) the population under study is a group of specific size with no new inputs (births or migration).

Depending on the type of model used on a case-by-case basis, each member of the population under study can—at any given time—be susceptible to an infection, become infected, or be immune. Susceptible individuals become infected upon contact with an already infected individual. As for infected individuals, they become immune at some point, whereas immune individuals retain their immunity.

A broad, yet simplistic classification of epidemic models distinguishes between the deterministic and stochastic types. In the former case, the focus is on what happens in a population on average, whereas the input parameters are fixed (for example the rate of disease or the rate of recovery) and thus this model makes predetermined predictions (for example the number of cases over time). On the other hand, stochastic models provide a range for each probability, such as the number of cases over time or the probability of occurrence of an epidemic.

The implementation of the model revealed that the death of 1,050 soldiers in 40 days meant that five new contacts of individuals had to take place every 4.5 days. In line with this finding, the study focused on determining which disease is characterized by this dynamic. Considering time intervals of 4.5 days as a time distribution model, scientists reached the conclusion that influenza could have produced such effects.[274] The estimated active contacts were then calculated for the total population of the city of Athens. Nevertheless, it is hard to estimate the toll of the epidemic on human lives. It is estimated that the population of Athens at the time—taking the arrival of refugees inside the city into consideration—ranged between 170,000 and 400,000 people.[275] For a city of 11 km² (including Piraeus and the area between the Long Walls and the port), even 170,000 people sufficed to cause overcrowding and poor health conditions.[276] Using a modification of the Reed-Frost formula, the ratio of five contacts applied to the hypothetical total of 400,000 people led to the finding that 100,000 people would have been infected within 40 days.[277]

The *Antonine Plague* (165/166 – 180 A.D. and 189 A.D.)

Another major epidemic that occurred in ancient times was the *Antonine Plague* (165/166 – 180 A.D. and 189 A.D.), which was named after the dynasty of Emperor Antoninus Pius (imp. 138 – 161 A.D.). According to the Roman historian Ammianus Marcel-

---

274 Anderson 1982; Morens 1992 and 1994.
275 Morens 1992 and 1994.
276 Morens 1992 and 1994.
277 Maia 1952; Morens 1992 and 1994.

linus (ca. 330—between 391 and 400), author of the last Latin history of the Roman Empire, the plague occurred in late 165 or early 166 A.D. on the eastern border of the Empire. This was the time when Lucius Verus (130 – 169 A.D.), the brother of the Emperor Marcus Aurelius (imp. 161– 180), was fighting the Parthians in the East together with his general, Avidius Cassius (ca. 130 – 175 A.D.).[278] Marcellinus believes that the disease originated from the desecration of the temple of Apollo in Seleucia and the ensuing transfer of his statue to Rome. But he also offers another explanation, namely that a soldier destroyed a tomb located at the base of the temple of Apollo, wherefrom sprang the scourge secretly kept there by "the innermost secrets of the Chaldeans" (... *ex adyto concluso a Chaldaeorum arcanis* ...).[279] Moreover, in the biography of Marcus Aurelius included in the *Historiae Augustae*, the epidemic is said to have originated in the land of the Parthians, wherefrom "the spirit of the plague (pestiferous vapor) escaped and then against the Parthians noxious completed the circle" (spiritus pestilens evasit atque inde Parthos orbemque complesse) and it spread throughout the world.[280]

Besides the supernatural approach of the Romans, it is believed that the disease was endemic in the Parthian region, having been transferred there from China via the Silk Road.[281] The first incidents occurred among the ranks of the Roman legions. The disease soon spread throughout the Eastern Mediterranean and Southeastern Europe. According to Ammianus Marcellinus, the epidemic was fatal and spread to France and possibly to the lower areas of the Rhine [282]:

> ... eiusdem Veri Marcique Antonini temporibus ab ipsis Persarum finibus ed usque Rhenum et Gallias, cuncta contagiis polluebat et mortibus ...

> ... in the same time Verus and Marcus Antoninus polluted everything with contagion and death, from the frontiers of Persia all the way to the Rhine and to Gaul ...

According to the *Historia Augusta*, the disease was brought to Rome by the army of general Avidius upon his triumphant return to the capital.

The *Antonine Plague* is often referred to as the *Plague of Galen*, after the famous Greek physician. We ought to clarify that Galen (129—after [?] 216 A.D.) indeed provides evidence of epidemics, relying on the Hippocratic concept of chymopathology (humoral pathology) in order to explain their nature, and gives a detailed clinical picture. None of his accounts, however, is used in the context of a single narrative concerning the particular epidemic, like the *Plague of Athens* by Thucydides, but rather his unsystematic accounts are scattered in different works, the contents of

---

278 Haas 2006.
279 Ammianus Marcellinus XXII.6 (ed. Gardthausen 1874: 327.22).
280 *Historia Augusta*, Verus VIII.2 (ed. Magie 1967: 222.2 – 3).
281 Fears 2004.
282 Ammianus Marcellinus XXIII.6 (ed. Gardthausen 1874: 326.10 – 12).

which refer to the pathogenesis of infectious diseases in general, as well as to their general clinical symptoms and signs.

As mentioned by Galen in his book describing the differences of various types of fever (Περὶ διαφορᾶς πυρετῶν—*On the Differences of Fevers*), a plague begins during a war or results from the vapors of swamps and lakes in the summer months. He even cites the case of the *Plague of Thucydides*, saying that it started in the houses due to the dirty air during the summer.[283] Moreover, the treatise *On Medical Definitions* (Ὅροι Ἰατρικοί) attributed to Galen deals with diseases in general on the basis of the type of fever that the patients suffer from, but in essence he explains the terminology of fevers. In fact, Galen thinks that the type of fever that he terms 'infectious fever' (*febris pestilens*) is common in all plagues characterized by symptoms of burning, vomiting, diarrhea, malodorous exhalation, catarrh, coughing, and ulceration of the trachea and the larynx.[284] During the *Antonine Plague*, death occurred nine to twelve days after the first symptoms appeared. In any case, the scholars who believe that the descriptions of Galen refer to the *Antonine Plague* have suggested smallpox (this view appears to be the most plausible), typhus fever, and plague as possible explanations, while others think that it is related to the *Plague of Athens*.[285]

Historically, the *Antonine Plague* coincided with both the Parthian War and the great invasion of the Marcomanni and Quadi (166–167 A.D.), who resided in modern Bohemia and Moravia and frequently invaded the northern territory of the Empire from 161 A.D. onwards. In 169 A.D., following the large Roman counter-offensive against the Marcomanni, the epidemic spread among the Germanic tribes. As a result the Romans found dead Germans scattered around on their way to the north.[286] Even in the ranks of the Roman army, however, the disease proved to be more dangerous than the Germans themselves, as it decimated the legions camped in the area of Aquileia.[287] In 180 A.D., in the midst of the Second Marcomannic War (176–180 A.D.), the Emperor Marcus Aurelius (imp. 161–180) died near Vindobona (modern Vienna), possibly because of the disease, but the Germans were unable to seize the opportunity to counter-attack as they had also been decimated by the disease. According to Galen, who moved to Aquileia by imperial order, numerous legionnaires died during the epidemic. The heavy losses forced the army to retreat towards Rome. In fact, Galen recounts that Marcus Aurelius was cured from a deadly abscess by praying to god Asclepius. He then carried on with his campaign, asking Galen to look after his son Commodus in the meantime.

After an absence of nine years, the disease reappeared in 189 A.D. at the time of the Emperor Commodus (imp. 180–192 A.D.), although its mortality must have been significantly lower and it possibly affected mainly the 'closed' populations of legion-

---

283 Galenus, *De differentiis febrium*, Liber I, Cap. VI (ed. Kühn 1824: 289–290)
284 [Galenus], *Definitiones medicae*, Cap. CXCIV (ed. Kühn 1830: 400).
285 Sabbatani 2009.
286 Cartwright 1972: 5–15.
287 Galenus, *De libris propriis*, 18–19 (ed. Mueller 1891: 99–100).

naires.[288] The population in the areas of the northern border of the Empire dropped sharply, as many citizens were forced to follow the Roman army in its retreat. Archaeological research has shown that the Roman power in the area declined, since the fortresses and cities along the defense line of Mannheim-Neckarau-Strasburg-Zurich-Bregenz-Regensburg-Passau were deserted for many years before the Roman legions returned to their old bases on the northern borders of the provinces of *Germania Superior, Raetia* and *Noricum.*[289]

The *Plague of Cyprian* (250 – 266 A.D.)

Concluding this brief account of the great epidemics of the ancient world, we ought to mention the interesting case of the *Plague of Cyprian*, which was named after the Archbishop of Carthage (ca. 200 – 258 A.D.). The plague appeared in 250 A.D., that is, sixty years after the previous *Antonine Plague* of 189 A.D. The *Plague of Cyprian* lasted until 266 A.D., leaving the Empire in a similarly dire situation as the *Antonine Plague*. According to Cyprian, the disease originated in Ethiopia, wherefrom it spread to the rest of North Africa and finally affected Rome. The clinical picture of this disease comprised diarrhea, vomiting, ulcerations of the nasopharynx, high fever, feeling of burning and thirst, generalized pustular rash on the trunk, and gangrene of the extremities. The disease spread throughout the Empire and even reached Hadrian's Wall in *Britannia Inferior*, affecting the Scottish tribes.[290] The emergency evacuation of cities contributed to the rapid spread of the disease, while the report of Cyprian reveals that the infectious agent was transmitted either directly from person to person or indirectly through objects.[291]

The *Plague of Cyprian* broke out during the most critical period of Roman history. Twenty-six emperors ascended to the throne between 235 – 285 A.D., and only one of them died of natural causes. Plague (250 – 266 A.D.) and the failed government of Valerian (imp. 253 – 260 A.D.) and his son Gallienus (co-imp. 253 – 260, imp. 260 – 268 A.D.) led the Empire to its greatest decline, while invasions on all its borders had become a common phenomenon. The Saxons were plundering Britain, Franks and Alemanni were invading Galatia, Marcomanni—the old enemies of the Romans—were sweeping the provinces south of the Danube, Goths and Sarmatians destroyed the eastern provinces, and the rising of Sassanids in Persia was threatening the weakened legions of Mesopotamia.

Summing up the data on the major epidemics of antiquity, we notice that they occurred and continued during periods of war (*Plague of the Achaeans*-Trojan War, *Plague of the Philistines*-War between Jews and the Philistines, *Plague of Athens*-Pe-

---

288 Galenus, *De libris propriis*, 18 – 19 (ed. Mueller 1891: 99 – 100).
289 Schönberger 1969.
290 Cartwright 1972: 5 – 15.
291 Cartwright 1972: 5 – 15. On the *Plague of Cyprian*, see more in Harper 2015.

loponnesian War, *Antonine Plague*-Parthian and Marcomannic Wars). An epidemic can occur in, and be limited to, a battlefield, like the plague that struck the Achaean camp, or spread to large urban centers like Rome, as was the case during the *Antonine Plague*. Sometimes an epidemic breaks out in a city, such as Ashdod, or spreads rapidly to an urban centre with existing public health issues due to overcrowding, as was the case in Ancient Athens. In any case, however, we ought to remember that the disease originated in Africa and the Middle East, which is particularly important for the evolution of a microorganism, as discussed in the chapters on ecology, environment and phylogeny.

Epidemics on the Borderline of the Ancient and Medieval Worlds

As we have already mentioned, for many scholars the search for, and identification of, reports on the main clinical symptom of plague, namely swollen lymph nodes, suffices to prove the existence of the disease in Europe since antiquity. The Greek term βουβών (meaning *groin*) essentially refers to the lymph nodes in the groin. The issue of buboes must be clarified.

With respect to this issue, the Byzantine physician of the fourth century A.D. Oribasius quoted a text by Rufus of Ephesus (first–second century A.D.) in his medical encyclopedias. According to Oribasius, Rufus seems to distinguish swollen buboes between those from accidental cause (found in the groin, armpits, and thighs) and those that occur with or without fever. Moreover, he refers to 'infected buboes' as a distinct species that occurs suddenly and is fatal.[292] Oribasius also describes the typical clinical picture of an infected bubo as described by Rufus: swollen and hard but without pus, painful and warm.[293] Even this description, however, covers a wide range of infectious diseases, including plague. In turn, Rufus cites Dioscorides and Posidonius in relation to 'infected buboes'.[294] At this point, it should be noted that the *Sixth Book of Epidemics* of the Hippocratic collection written in the late 5th or early 4th century B.C. mentions buboes, but it does not associate them with any infectious disease or plague in particular.[295] Apparently, all references to buboes up to the Early Middle Ages did not indicate a specific disease, but rather a general set of symptoms of specific infectious diseases. Similarly, any references

---

292 Oribasius, *Collectiones medicae*, Lib. XLIV.14.1 Ἐκ τῶν Ῥούφου. Περὶ βουβῶνος (ed. Raeder 1931: 131–132).
293 Oribasius, *Collectiones medicae*, Lib. XLIV.14.1 Ἐκ τῶν Ῥούφου. Περὶ βουβῶνος (ed. Raeder 1931: 131–132).
294 Oribasius, *Collectiones medicae*, Lib. XLIV.14.2 Ἐκ τῶν Ῥούφου. Περὶ βουβῶνος (ed. Raeder 1931: 132.8).
295 *Corpus Hippocraticum*, *De morbis popularibus VI*, 2 (ed. Littré 1839–1861: 5.388 Greek text and French translation wit note 3; see also pp. 57, 60, 70, Littré's introduction with commentary on all the passages in the books *De morbis popularibus II, IV, V, VI, VII*, referring to buboes).

to 'buboes' did not necessarily point to plague. For the physicians of that time, a 'bubo' was the typical reaction of the organism to a disease; it cannot be used for the differential diagnosis of a particular disease. On the other hand, when not anatomically defined, the term 'bubo' may refer to the swelling of any group of lymph nodes (in the armpits, or in the neck, for example) and not necessarily to the lymph nodes in the groin. Based on selective data, numerous researchers support the view that plague had existed in Europe for centuries.

All diseases mentioned above, however, are merely classical examples of the susceptibility of human populations in antiquity to emerging diseases, which explains their great mortality rate and spread. Before the division of the Empire, the ancient historian Eusebius of Caesarea (263–339 A.D) reported an epidemic that struck the eastern part of the Empire in the winter of the year 312–313.[296] The epidemiological fact he mentioned is related to winter rains and the emergence of a disease. The disease was characterized by generalized pustular rash and loss of vision. Apparently, the necrotizing inflammation of the skin and subcutaneous tissue was interpreted as carbon, a term provided by Eusebius himself. The mortality rate was high in both the urban centers and the countryside, a fact that also suggests the theory of smallpox. During the year 333 and according to Eusebius and the *Chronographia* of the later Byzantine chronicler Theophanes (758–817 A.D.), a disease broke out in Syria, Antioch, and Cilicia as a result of prolonged food shortages.[297] Between March and June 346, an epidemic broke out in three monasteries in the area of Thebes, Egypt, resulting in the death of 100–300 monks.[298] Among the various symptoms of the disease, the redness of the eyes and asphyxia of patients are of particular interest, as they may point to pharyngeal or laryngeal diphtheria.[299] Notably, the epidemic reemerged in Thebes and its monasteries in 360–361.[300] In 359, an epidemic was also reported in the Byzantine city of Amida, which was then besieged by the Persians, with countless dead people lying on the streets of the city. Likewise, the Byzantine military units on the border with Persia were ravaged four years later by an unknown epidemic, according to another Byzantine historian, Philostorgius (368–439 A.D.).[301] But, even in the first decades of the existence of the Eastern Roman Empire, numerous epidemics were reported although their symptoms have not been described, such as the epidemics in the province of Macedonia (383), Antioch (384–385), and Palestine (406).[302] As a letter written around 405 (or 406–407) by Ioannis Chrysostomos (c. 347–407) to Pope Innocent I (reign. 401–417 A.D.) reveals, Asia

---

296 Eusebius, *Historia Ecclesiastica*, 9.8 (ed. Migne 1857: 815–819).

297 Eusebius, *Hieronymi Chronicon*, CCLXXVIII [= 335] CCLXXVIII Olymp. [= A.C. 333] (ed. Helm 1913: 233, e); Theophanes, *Chronographia*, A.M. 5824 [= A.C. 333], P. 23 (ed. Classen 1839: 42.16–43.6).

298 As reported by Stathakopoulos 2004: 185–191.

299 Chen-Chia 1997.

300 Stathakopoulos 2004: 185–191.

301 Philostorgius, *Historia Ecclesiastica*, 5.2 (ed. Bidez 1913: 67–68).

302 Stathakopoulos 2004: 185–191.

Minor was struck by an epidemic amid war and the siege of cities, but these cities are not specified [303]:

> ... ἐπεὶ καὶ ἡμᾶς τρίτον ἔτος τοῦ τὸ ἐν ἐξορίᾳ διατρίβοντας, λιμῷ, λοιμῷ, πολεμίοις, πολιορκίαις συνεχέσιν, ἐρημίᾳ ἀφάτῳ, θανάτῳ καθημερινῷ, μαχαίραις Ἰσαυρικαῖς ἐκδεδομένους ...

> ... third year in exile, suffering from famine, plague, war, continuous sieges, desolation and daily deaths, remaining exposed to the swords of the Isaurians ...

During the fifth century, numerous cities and regions were affected by epidemics: Antioch in 440, Constantinople between 445 and 447, and Asia Minor (Phrygia, Galatia, Cappadocia, Cilicia) between 451 and 454, while the capital was once again struck by famine and plague in the period 452–458.[304] Unfortunately, in all these cases no direct or indirect data exist with regard to the nature of the disease. On the eastern border of the Empire, Edessa, Antioch, and Nisibis (modern Nusaybin, in Turkey) were also ushered in the sixth century amid epidemics.[305] The West did not remain unaffected by the epidemics, as during the siege of Rome by the Goths in 537, the people and the army of the Byzantine general Belisarius (500–565 A.D.) were faced with a harsh epidemic that broke out in the city. In 539 and in the midst of the Gothic Wars, epidemics occurred in Emilia and Tuscany.[306] In that same year, the Goths and Franks conquered Mediolanum (modern Milano), thus changing the balance of the war in northern Italy. The Franks, however, overestimated their power and, except for the Byzantines, turned against their former allies, the Goths; but a major epidemic that broke out in the Po Valley devastated their army and forced them to escape hurriedly across the Alps.[307]

The non-exhaustive reports on epidemics during the transition period from Late Antiquity to the Early Middle Ages may be due to the fact that epidemics were mere outbreaks of endemic diseases, which were regarded as normal by contemporary people. Similarly, Byzantine scholars soon became acquainted with plague and over the next centuries they simply use the word *thanatikon* to describe the terrible disease.

Based on the above and taking into account the lack of relevant mentions in available sources, it appears that the possibility of the existence of plague before the outbreak of the first pandemic is negligible, unless the emergence of some new historical source proves otherwise. No clear indication suggests the existence of the disease; on the other hand, numerous sources describe plague since its first emergence. The surprise resulting from the disease that was bound to torment the Empire was recounted with various degrees of accuracy. But the eschatological na-

---

303 Stathakopoulos 2004: 221–229.
304 Stathakopoulos 2004: 231–241.
305 Stathakopoulos 2004: 250–255.
306 Stathakopoulos 2004: 272–274.
307 Browning 1978: 175–177.

ture of these texts is obvious, as they associate high mortality with Divine wrath. If plague had appeared before 542, it would have definitely caused the same damage as it did in the sixth century, staggering contemporary scholars who would not have limited themselves to a mere mention of the kind of Eusebius' account, who overlooks the event and simply informs us about an epidemic in the East.

Finally, the use of the word 'plague' in modern medico-historical studies to describe the cases mentioned above should perhaps be replaced by another term. Maybe the 'neutral' term *epidemic* is more appropriate (*Epidemic of Athens*, the *Antonine Epidemic*, for example), as it does not generate misunderstandings and is not related to the official medical term 'plague' that denotes the disease itself.

Chapter 5
First Plague Pandemic

Chronicle of the *Plague of Justinian*

One of the main sources of information regarding the events that took place during the *Plague of Justinian* is the *History of the Wars* by Procopius, more specifically the book on the Persian War.[308] Procopius served as advisor and military secretary to the General Belisarius, and he accompanied him on his victorious campaigns against the Persians, Vandals, and Goths. He was in direct contact with the imperial couple Justinian (reign. 527–565) and Theodora (ca. 500–548), and he was familiar with the secrets of the palace, in addition of being an eyewitness to the plague.[309]

According to Procopius' narrative, the epidemic started in Pelusium of Egypt. Procopius indirectly points to the exact year of the outbreak of the epidemic, as the narrative is included in the account of the military operations of the Persian War in the summer of 542. More specifically, he reported that the plague broke out during the thirteenth year of Justinian's reign, and that the disease reached Constantinople within two years after its onset in Pelusium, perhaps in mid-July.[310] Procopius points out that the disease spread throughout the globe and seemed to be characterized by a certain periodicity. His view that, wherever the disease appeared, it became endemic for some time, is interesting. Then it spread to the neighboring areas or returned with the same intensity to the areas it had already affected.

From Pelusium, the disease spread towards Alexandria and gradually affected the rest of Egypt before moving eastwards to Palestine. It reached Alexandria perhaps in mid-September 541, and harshly affected the residents of the city and the more so its poorer social strata.[311] Alexandria was deserted, and many residents fled to other cities as the plague spread throughout Egypt.

The disease moved eastwards to Palestine at about the same period. A study of the memorial inscriptions in the cemeteries of the area reveals increased mortality rates in Gaza, a fact that could be linked to the outbreak of plague in Palestine.[312] Presumably, the disease reached Gaza, Ashkelon, and the cities of the Negev region through the trade routes of the area, approximately one month after its onset in Pelusium, that is, in mid-August.[313] The disease moved along the coastal cities and then reached the inland of Palestine, apparently through the trade routes; it struck Jeru-

---

308 Procopius, *De Bellis, De Bello Persico*, II.22.7–23.12 (ed. Haury 1905: 249.7–259.25).
309 Nicol 1991: 108.
310 Stathakopoulos 2004: 99–106, 277–278.
311 Stathakopoulos 2004: 280; Michael Syrianus, *Chronica*, Liber IX, Cap. XXIX (ed. Chabot 1899–1910: 4.309–310; Fr. transl. Chabot 1899–1910: 2.241–248, esp. 244).
312 Glucker 1987: 124–127.
313 Kislinger 1999.

https://doi.org/10.1515/9783110613636-008

salem and its surrounding areas in 542. In spring 542 it moved westwards to Syria, while, in the same year, it appeared in Antioch, Apamea, Epiphanea, and Emesa (modern Homs in Syria), but it is impossible to tell with certainty which of these cities was affected first.[314]

The key city in the pathway of plague towards the Byzantine capital may have been Myra on the southwestern coast of Asia Minor, which was a stop along the sailing from Alexandria to Constantinople.[315] The transport of Egyptian grain, which was directly linked to the fluctuations of the Nile, lasted between 22 and 26 days. According to Procopius, a ship could make up to three trips (Alexandria-Constantinople) before the winter (we suppose during the time between April and October).[316] Based on this assumption, we may infer that plague reached Myra in mid-spring of 542. It struck the city with vehemence.

The disease reached Constantinople in 542, and it soon spread throughout the city. According to Procopius' account, the epidemic lasted four months, from April through August. The early symptoms were common to all patients. Low fever was common and did not worry either the patients or the physicians. Within 24 hours or a few days, the general situation of the patients rapidly declined, and lymphadenitis in the axillary and inguinal lymph nodes appeared. According to Procopius, the disease progressed differently among different patients, as it could cause coma or a form of insanity accompanied by insomnia, delusions, and hallucinations. The latter two are responsible for a large number of deaths, as many patients believed that they could hear the voices of spirits calling them and thought they were persecuted by these spirits. In that state of delirium, many patients jumped into wells or off the city walls in order to escape from their hallucinations. The information regarding the rates of maternal and neonatal mortality during the plague epidemic is particularly interesting. It appears that mortality was extremely high: according to Procopius, only three pregnant women survived even though they had a miscarriage, whereas only one infant survived after birth. Procopius also reports that the disease was not transmissible from person to person, since those in close contact with patients (relatives and physicians) did not become infected. New patients were not necessarily members of the same family, and everyone could become infected anywhere.

An interesting element in Procopius' account is the 'ignorance' of physicians and the random semantic approach to the disease. Being ignorant of the nature of the disease, the physicians believed that buboes were at the epicenter of the disease. The necropsies that they carried out revealed the familiar 'anthrax' of the buboes. When carefully studying the account of Procopius, one may wonder how a situation present for already quite some time, namely the epidemic that slowly spread from Egypt to the south-eastern provinces, caught by surprise the residents and physicians

---

314 *Vita Sancti Symeonis Stylitae*, chapters 59–60 (ed. van den Ven 1962–1970: 1.52–54); Evagrius, *Historia Ecclesiastica,* IV 29 (eds Bidez and Parmentier 1898: 177–179).
315 Zimmerman 1992.
316 Procopius, *De aedificiis,*V.1.19–10 (ed. Haury 1913: 150.25–151.7).

of Constantinople when it eventually reached the city. Many patients died immediately, while others passed away after many days. Those who died immediately had hematemesis and a black vesicular rash.

Procopius presents a grim picture of the number of victims, arguing that as many as 5,000 deaths occurred daily at the peak of the disease. John of Ephesus (ca. 508-ca. 588 A.D.) even reported 16,000 deaths on a daily basis, while the *Chronographia* of Theodosius (10th cent.) includes a phrase that was typical of the chronographers of the subsequent centuries, stating [317]:

> ... μένειν ἀτάφους τοὺς τεθνεῶτας διὰ τὸ μὴν ἐξαρκεῖν τοὺς κραββάτους τῶν ἐκκλησιῶν καὶ τῶν οἴκων, ἐκφέρειν αὐτούς ...

> ... the deceased were left unburied since there were not enough stretches to transfer them from the churches and their home ...

The problem of irregular burial and abandonment of corpses was settled by Justinian's order to the *referendarius* Theodoros (the official who reported the citizens' demands to the emperor). With the help of a body of volunteers, Theodoros took care of digging graves. But it soon became apparent that the number of deceased was excessively high. Theodoros was forced to stack the bodies of the deceased in the towers of the walls. Procopius even mentions the intense odor that emerged from the towers, which horrified and scared the inhabitants whenever the wind blew from that area to the inner city. Finally, Procopius informs us that the epidemic spread to the Persians, without making it possible for us to understand when exactly the incident took place. Maybe he refers to the outbreak of the epidemic in Syria in spring 542 or to some other epidemic of which Procopius was later informed.

During the epidemic, "it so happened the the Emperor (Justinian) became ill, his buboes too had been swollen" (ἄλλως τε ἡνίκα βασιλεῖ νοσῆσαι ξυνέβη καὶ αὐτῷ γὰρ ξυνέπεσε βουβῶνα ἐπῆρθαι). According to Procopius, Justinian was eventually cured through the intercession of the saints Kosmas and Damianos (the so-called Ἅγιοι Ἀνάργυροι [*Agioi Anargyroi*], *the Holy Unmercenaries*), who saved him from the paradox and absurdity (delirium?), and helped him get back on his feet.[318] The news of Justinian's illness had deep political implications and created a major problem as to his succession to the throne. Empress Theodora assumed the rule of the Empire, and, despite widespread antipathy from the people and the ranks of the army, she was eventually acknowledged for successfully handling the burning issues facing the Empire.

The general narrative of Procopius reveals the surprise of everyone in the capital in the face of the disease, while it appears that the events exceeded the ability of the state to respond to the situation. At a time when the microbial nature of the disease, its mode of transmission, and even the concept of quarantine were still unknown,

---

317 Theodosius Melitenus, *Chronographia, Justinianus*, 8–10 (ed. Tafel 1859: 89).
318 Procopius, *De aedificiis*,I.6.5 (ed. Haury 1913: 30.2–9).

existing accounts show that Constantinople was caught completely by surprise. At that time, information was definitely conveyed at a different speed than nowadays, but, given that the epidemic had already been raging in the Byzantine provinces for nine months, it seems reasonable to ask the following questions: a) Was the well-organized Byzantine state apparatus fully aware and informed of the evolution of the epidemic in the provinces? b) Did the authorities in Constantinople consider it to be another 'typical' epidemic of an endemic disease that was bound to end soon? c) If the authorities were indeed properly informed of the situation in Egypt, did they realize that the disease had nothing to do with what they had ever dealt with by that time? d) If the authorities knew about the disease, why did nobody believe that it could eventually reach Constantinople? All these questions have no definitive answer and only speculations can be made.

While the epidemic was raging in Constantinople, however, the disease also appeared in central Asia Minor and more specifically in the regions of Ankyra (now Ankara) and Galatia.[319] Given that Ankyra was an interim station in the road network that connected Constantinople and Antioch (through the provinces of Bithynia, Galatia, Cappadocia, and Cilicia), two scenarios seem plausible.[320] The disease may have reached the inland of Asia Minor through Constantinople (which is a more possible event) during the epidemic that had broken out in the capital, or through Syria which was also affected by plague at the time. The devastating year of 542 seems to have ended with the disease spreading to the coasts of North Africa and Sicily.[321]

The following year of 543 was characterized by the gradual appearance of the disease in other provinces of the Byzantine Empire, in addition to its reappearance in eastern provinces that had already been affected. It may have spread to Illyria due to the movement of the Byzantine army to Italy during the resurgence of the Gothic War (542–552).[322] When the praetorian prefect (*praefectus praetorio*) Maximinus was ordered by Justinian to confront the Goth King Totila (reign. 541–552) who had recaptured almost all of Italy, he camped in Epirus for some time before disembarking on the Italian coasts. The Byzantines planned on landing in Sicily and simultaneously attacking the Goths in the south-west with the recapture of Naples by the *magister militum* (commander) Demetrius. It is worth noting that the appearance of the Byzantine fleet in Syracuse may have coincided with the disease in Sicily in late 542.[323] But maybe the failure of the Byzantines to land in southern Italy was the result of the epidemic that broke out in Illyria in 543, and not of the incompetence of Maximinus—as he had been accused at the time. While this situation was unfolding in Illy-

---

319 *Vita Theodori Sycionis*, 8 (ed. Festugière 1970: 1.7 Greek text, 2.8 French translation).
320 Cuntz 1929: 91.
321 Cited from Stathakopoulos 2004: 290.
322 Marcellinus Comes, *Chronicon*, 543.2, *VII Post Consulatum Basili II Anno III* (ed. Mommsen 1894: 107.2).
323 Stathakopoulos 2004: 290–293.

ria, the disease appeared on the coast of North Africa, in modern Algeria, slowly surrounding the Mediterranean Sea.[324]

According to the Syriac *Chronicle* attributed to Dionysios of Tel-Mahre (d. after 775), the disease struck Mesopotamia between 546 and 547. As mentioned by the Pseudo-Dionysios text, madness and rage broke out in areas of Mesopotamia during the period 855 A.G. (*Anno Graecorum*, Seleucid era system of numbering years) and 858 A.G. (the years of the reign of Justinian, 543/44 – 546/547).[325] The next reference to plague was recorded around 557 – 558 in Amida, at the border of the districts of Mesopotamia and Armenia, when 35,000 residents died within three months.[326]

The Byzantine chroniclers Agathias (ca. 530 – 582 or 594 A.D.) and Ioannes Malalas (ca. 491 – 578) cite testimonies for the epidemic of 558 in Constantinople.[327] According to these testimonies, the nature of the disease was the same as in 542 [328]:

... ἡ λοιμώδης νόσος αὖθις τῇ πόλει ἐνέπεσε καὶ μυρία διέφθειρε πλήθη... ἡ δὲ τῆς νόσου ἰδέα παραπλησία τῇ πρεσβυτέρᾳ ἐτύγχανεν οὖσα ...

... the infectious disease once again befell the city and killed the thousands ... the image of the disease was similar to [that of] the older one ...

Just like Procopius, Agathias mentions cases of sudden deaths and whole families who vanished. He finally provides some interesting information, like the fact that the disease mainly affected children, adolescents, and adult males, while the female population suffered less. According to Malalas, the epidemic broke out in February and lasted six months, and the *Chronographia* of Theophanes includes the same reference (February–July).[329]

A plague epidemic struck Anazarbus and Antioch in 560 – 561, in the neighboring provinces of Cilicia and Syria, respectively, with high mortality rates among the residents of those areas.[330] Fifteen years after the second epidemic of Constantinople and during the eighth year of the reign of Justin II (565 – 578), a third epidemic broke out around 573 – 574 which seems to have struck the city with vehemence. Sources speak of 3,000 deaths per day. Prior to the epidemic a strong earthquake had hit Con-

---

324 Stathakopoulos 2004: 290 – 293.
325 On the episode of 855 A.G. = 543/544, see Pseudo-Dionysios, Chronicon, 855 and also 858 (ed. Chabot 1933: 119 and 112, respectively; Eng. transl.: Witakowski 1996: 107 and 102, respectively). For its source (Johannes Ephesinus, Historia Ecclesiastica), see Witakowski 1991: 265 (on 855 A.G.) and 266 (on 855 and 858 A.G.).
326 On the episode of 869 A.G. = 557 – 558, see Pseudo-Dionysios, Chronicon, 869 (ed. Chabot 1933: 119; Eng. transl.: Witakowski 1996: 107). For its source (Johannes Ephesinus, Historia Ecclesiastica), see Witakowski 1991: 266).
327 Agathias Myrinaeus, *Historiae*, V.10, A.C. 558, P. 153 – 154 (ed. Niebuhr 1828: 297.21 – 299.14); Ioannes Malalas, *Chronographia*, LXVIII, O. 232 – 235 (ed. Dindorf 1831: 488.4 – 489.18).
328 Agathias Myrinaeus, *Historiae*, V.10, A.C. 558, P. 153 – 154 (ed. Niebuhr 1828: 297.21 – 299.14)
329 Ioannes Malalas, *Chronographia*, LXVIII, O. 232 – 235 (ed. Dindorf 1831: 488.4 – 489.18).
330 Theophanes, *Chronographia*, A.M. 6053 [= A.C. 553], P. 199 (ed. Classen 1839: 364.3 – 6).

stantinople plunging its residents into despair.[331] Antioch evidenced a similar situation in 580–581, when a destructive earthquake was followed by an epidemic.[332] In the year 585–586, and during the fourth year of the reign of Emperor Maurice (582–602), a new epidemic broke out in Constantinople.[333] The report of 400,000 deaths resulting from the epidemic obviously lacks credibility. In the late sixth century, Antioch—the second most preferred city of plague—was once again affected (591–592), as reported by eyewitness Evagrius (536–594 A.D.).[334] Evagrius recounts that plague first appeared in Antioch in 542, when he was still a child. In fact, he had been infected, but he was able to survive although plague killed many members of his family. The most interesting aspect of his account beyond the typical symptomatology concerns the periodicity of the disease, which appeared in Antioch every second year of each new 15-year period, a fact that is not far from reality[335]:

> ... ἤδη τετράκις ἐπισκήψαντος τοῦ πάθους ἀνὰ τὴν Ἀντιόχου ἐπεὶ τέταρτος ἀπ' ἀρχῆς διῆλθε κύκλος ... πανωλεθρία δὲ σχεδὸν συνέβαινε τοῖς ἀνθρώποις ἐπέπιπτεν ἐν τῷ δευτέρῳ ἔτει τῆς πεντεκαιδεκαετηρίδος τοῦ κύκλου ...

> ... the disease has so far affected Antioch four times upon completion of the fourth cycle since the first time (outbreak) ...the calamity fell upon the people every second year of the fifteen-year cycle ...

During the same period in the West, Paul the Deacon (ca. 720—ca. 799) recounts an epidemic in the Exarchate of Ravenna and the regions of Istria and Grado. The Exarchate was a political creation of the Byzantine Emperor Maurice, who assigned supreme administrative and military power to an 'exarch' as a surrogate of the emperor in the remains of the former Byzantine territories in Italy. Paul recounts that this epidemic was transferred through the sea; apparently it had broken out in Rome and the rest of central Italy. He characteristically recalls that it was similar to the epidemic that had broken 30 years earlier, obviously referring to the epidemic of 565, the epicenter of which had once again been Rome.[336] The Exarchate of Ravenna was struck by plague again in 600.[337]

Another major city affected by plague was Thessaloniki. It was struck by an epidemic in the summer of 597 as the *Miracula Sancti Demetrii* (*Miracles of Saint Deme-*

---

331 Michael Syrianus, *Chronica*, Liber X, Cap. XIX (ed. Chabot 1899–1910: 4.373–374; Fr. transl. Chabot 1899–1910: 2.349–353, esp. 352).
332 Agapius, *Historia universalis*, 178 (ed. Vasiliev 1912: 438).
333 Agapius, *Historia universalis*, 179 (ed. Vasiliev 1912: 439).
334 Evagrius, *Historia Ecclesiastica*, IV.29 (eds Bidez and Parmentier 1898: 178.11–16).
335 Evagrius, *Historia Ecclesiastica*, IV.29 (eds Bidez and Parmentier 1898: 178.11–16).
336 Paulus Diaconus, *Historia Langobardorum*, II.4 (eds Bethmann and Waitz 1878: 74), with the following title in the table of chapters: "De signis pestilentiae et mortalitate quae tempore Narsetis Italiam vastavit".
337 Paulus Diaconus, *Historia Langobardorum*, IV.14 (eds Bethmann and Waitz 1878: 121), with the following title in the table of chapters: "De peste aput Ravennam et mortalitate aput Veronam".

trios) tell us.[338] The narrative refers to a deadly epidemic that hit the city and its out-skirts with a symptomatology indicative of bubonic plague. According to the mira-cles, Saint Demetrios saved the city from the devastation of plague, as he had done in the past when he saved the city during various sieges.

The end of the 6th century is characterized by a resurgence. In 598, a plague epi-demic broke out in Drizipera (modern Büyük Karıştiran in European Turkey) among the ranks of the army of the Avars who were threatening Constantinople.[339] The out-break of the disease was enthusiastically welcomed by the Byzantines, as it decimat-ed the Avars; it was seen as their punishment for the desecration of the tomb of Alexander Martyr by the barbarians. Nevertheless, the disease reappeared in the fol-lowing year in Constantinople and Bithynia.[340] The later *Chronicle* of the Jacobite pat-riarch of Antioch Michael the Syrian (1126–1199) speaks of a plague epidemic with an incredibly high number of victims, among them the Patriarch John of Constanti-nople. Michael believed that the epidemic originated in Bithynia, wherefrom it spread throughout Asia Minor. Given that a source from the following year refers to the epidemic of the Avars in Drizipera, in eastern Thrace, it is very likely that this was the continuation of the same epidemic that struck Constantinople and its surrounding provinces. The sixth century ended with a new epidemic in Syria in 599, which is possibly associated with the aforementioned spread of an epidemic or-iginated in Asia Minor.[341]

While the Empire was going through a period of political stability during the reign of Emperor Heraclius, Constantinople was hit by a new catastrophe in 618. This epidemic is fragmentarily described in various sources, the combination of which allows us to draw a rough picture of the events. The *Miracles of Saint Artemios* ought to be mentioned as a typical case of the difficult interpretation of information and its possible misinterpretation. They narrate the history of a twelve-year-old girl named Euphemia, who became ill during a plague epidemic that provoked the death of many people at the time of the reign of Heraclius (in 618?). Euphemia suf-fered from a 'bubonic swelling' of the axillary lymph nodes and was moribund. She was speechless for two days, and her body was full of black spots (a vesicular rash, apparently) which were seen as a 'blessing' by her fellow citizens, but Saint Artemios saved her.[342] Plague is also considered to have been the epidemic described by Nike-phoros the Patriarch of Constantinople (d. 828 A.D.) in his *Short History* (*Breviarium*) of the years 602–729. At that time, the Persians had taken Egypt from the Byzantines. Trade of grain from the former granary of the Empire had become impossible, and

---

338 Lemerle 1979: 57–82.
339 Theophylactus Simocatta, *Historiae*, VII.15 (ed. Bekker 1834: 297–298).
340 Michael Syrianus, *Chronica*, Liber X, Cap. XXIII (ed. Chabot 1899–1910: 4.386–387; Fr. Transl. Chabot 1899–1910: 2.371–374, esp. 373–374).
341 Conrad 1994: 17–26.
342 *Miracula Sancti Artemii*, 52 (ed. Papadopoulos-Kerameus 1909: 34).

Byzantium was struck by plague and ravaged by death.[343] Here, however, the *Miracles of Saint Artemios* appear to focus more on the importance of faith rather than on the actual depiction of the events of the epidemic. The existence of plague does not imply that other infectious diseases were absent. In the case of Euphemia, the reference to a 'blessing' leads our diagnosis towards smallpox.

The epidemic that exploded in Alexandria between 618 and 619 is maybe related to the aforementioned epidemic in Constantinople. There is a temporal gap, however, since the epidemic of Constantinople seems to have taken place after the conquest of Egypt, whereas, according to *Life of Saint John* (John V of Alexandria [ca. 550–620]) by Leontius of Neapolis (590–668 A.D.), the epidemic broke out in besieged Alexandria before its conquest by the Persians.[344] If a chronological relationship exists, the epidemic could be related to the rest of Egypt, which did not receive grain due to the Persian presence. The siege of Alexandria lasted for many months and it took around three years for Egypt to surrender (approximately in 620).[345] Unfortunately, the symptoms of the epidemic are not described. Plague may not have been involved after all, since the city was under siege for a long time and dysentery seems also plausible.

In the late seventh century Constantinople was once again hit by an epidemic, according to the *Chronographia* of Theophanes. During the dredging works at the Neorion harbor on the southern shore of the Golden Horn in 698, large volumes of waste were removed resulting in a plague that struck the city [346]:

... Νεωρήσιον λιμένα ἐκκαθαίροντος ἡ τοῦ βουβῶνος λύμη ἐνέσκηψε τῇ πόλει ...

... while cleaning the port of Neorion the impurity of the bubo appeared in the city ...

The Patriarch Nikephoros, who describes the events surrounding the dethronement and mutilation (nose slitting) of the Emperor Leontius (imp. 695–698), speaks of an epidemic that decimated the population [347]:

... νόσος λοιμικὴ τῇ πόλη ἐπέσκηψε καὶ πλῆθος λαοῦ ἐν μησὶ τέτρασι διέφθειρεν ...

... an infectious disease befell the city and killed numerous people within four months ...

But maybe Nikephoros refers to the outbreak of the epidemic recorded by Theophanes. An epidemic broke out in Crete during the eighth century, but no data exist as to its duration and mortality rate (probably between 711–740).[348]

---

343 Nicephorus Constantinopoli Patriarchus, *Breviarium*, 8.2–7, and 12.7–9 (ed. and transl. Mango 1990: 48–49, and 54–55).
344 Leontius, *Vita Sancti Joannis*, 24 (ed. Festugière 1974: 375).
345 Stratos 1978: 229–230.
346 Theophanes, *Chronographia*, A.M. 6190 [= A.C. 698], P. 309–310 (ed. Classen 1839: 568.1–3).
347 Nicephorus Constantinopoli Patriarchus, *Breviarium*, 41.23–24 (ed. and transl. Mango 1990: 98–99).
348 Detorakis 1970-71.

In the summer months of 718 an epidemic raged in Constantinople and its sub-urbs, which coincided with the siege by the Arabs. The report mentions the famine of the Arab army and the epidemic that broke out in the besieged city, perhaps as a re-sult of the severe health conditions. The statement that the disease killed 300,000 is certainly excessive.[349] If mortality was high, however, this could possibly be regarded as one of the last occurrences of the disease in the city.[350] According to the *Cronicle of Zuqnīn*, plague appeared in Syria, Mesopotamia and Iraq between 744 and 745. [351] The *Chronicle* attributed to Dionysios is a Syriac chronicle reporting the historical events from the biblical Genesis to the years 775–776 A.D. It was probably compiled in the monastery of Zuqnīn near Amida.

Between 745 and 746 plague appeared in Greece, although the areas that were affected remain unknown. The only precise place name in the Greek territory that is mentioned is Monemvasia in the Peloponnese.[352]

The first plague pandemic in the Byzantine Empire came to an end in 747–748, when it appeared in Constantinople for the last time. Its intensity was similar to that of the first epidemic 200 years earlier, in 542. Presumably, this was the evolution of the epidemic of 745, which continued until the summer of 748 and affected regions outside the Empire as well. According to the sources, the destructive disease struck the city and wiped out all residents.[353] Nikephoros pays particular attention to this event and his account reminds us of Procopius, who describes patients' delirium and the ghosts that appeared suddenly and led people to behave strangely (like fight-ing against each other with their swords).[354] The reports speak of a high case fatality rate, while residents were forced to use the gardens of their homes as burial grounds since the cemeteries were no longer enough for the infinite number of dead bodies. Just like in the year 542 when the dead were stacked in the towers, the dry water tanks were turned into makeshift mass graves. According to Nikephoros, the epidem-ic lasted for a year before it gradually disappeared.[355] The consequent demographic catastrophe seems to have been significant, judging from the fact that, in 754–755, the Emperor Constantine V Kopronymos (imp. 741–775 A.D.) ordered people to move from various regions of Greece to the capital in order to restore the city to its pre-epidemic situation. The era of the epidemic coincided with a generally trou-bled period in the history of Byzantium (Iconoclasm), and the terrible epidemic was considered as divine punishment upon the iconoclast emperor for his sins.

---

349 Rochow 1991: 92.
350 Theophanes, *Chronographia,* A.M. 6209 [= A.C. 718], P. 333 (ed. Classen 1839: 611.8–10).
351 *Chronica Zuqnīni, ad a. 1055–56* (ed. Harak 1999: 168–174).
352 Theophanes, *Chronographia,* A.M. 6238 [= A.C.745], P. 354 (ed. Classen 1839: 651.14–18).
353 *Chronica Zuqnīni, ad a. 1055–56* (ed. Harak 1999: 168–174).
354 Nicephorus Constantinopoli Patriarchus, *Breviarium,* 67.5–40 (ed. and transl. Mango 1990: 138–141).
355 Nicephorus Constantinopoli Patriarchus, *Breviarium,* 67.5–40 (ed. and transl. Mango 1990: 138–141).

These written testimonies are essentially the last to mention a bubonic plague epidemic within the ever changing borders of the Byzantine Empire during the eighth century. Historical sources remained silent until plague resurfaced in Constantinople as the *Black Death*, six centuries later.

Bubonic Plague in the Old Territories of the Byzantine Empire

The seventh century was one of the most critical in the history of Byzantium, since numerous events of paramount historical importance took place. At the dawn of this century the Emperor Maurice (imp. 582–602) had been murdered, and the Empire went through the tyranny of Phocas between 602 and 610. In 610, however, Heraclius, whose name was linked with significant milestones in European history, ascended to the throne (610–642). The Byzantine Empire was never going to be the same again. The results of the events of the period 629–642 were disastrous for the Empire, whereas the basis for the creation and rise of a new force was laid, that is, Islam. This development forces us to study the next historically recorded plague epidemics based on the territorial changes that took place in the Empire. In the following chapters of this study, the epidemics that took place in the Byzantine provinces that had been lost will be summed up in the context of the historical events that contributed to the loss of these areas.

Italy

At the time of Justinian, the formerly ecumenical Roman Empire was limited to the areas of Eastern Mediterranean, as the western part had collapsed under the weight of the major movements of populations and their raids (Ostrogoths, Visigoths, and Vandals). Italy, which had been part of the Ostrogothic Kingdom, was recaptured by the Byzantines during the great Gothic War (535–540). The Goths rebuilt their forces and the war resurged (542–552), but the Byzantines eventually consolidated their domination in 561–562.[356] Nevertheless, the Byzantine presence in Italy turned out to be short-lived, and the advent of the Lombards (or Longobards) fragmented and gradually limited the imperial possessions to the Exarchate of Ravenna.[357] The fall of Ravenna in 751 marked the end of the Byzantine presence in the Italian Peninsula, with the exception of a brief glimmer of hope in mid-eleventh century.[358]

We can extract indirect information about the first outbreak of plague in Italy from the *Inscriptiones Christianae Urbis Romae* collection, which includes the memo-

---

356 Moorhead 2008.
357 Moorhead 2008.
358 Brown 2008: 437–444.

rial inscriptions from the cemeteries of Rome. Its study reveals an increased number of inscriptions between November 543 and February 544. This increased number possibly coincides with the appearance of plague in Italy, which allegedly came from North Africa.[359] We cannot exclude, however, the possibility that the disease was initially brought to Sicily by the Byzantine fleet, wherefrom it spread to the rest of Italy during the campaign against the Goths in 543. Another epidemic was recorded in Rome and its surrounding areas between 554 and 555, which mainly affected the child population.[360] The next outbreak of plague on the Italian Peninsula dates to the period between 565 and 571. It spread across the Italian north, and it also affected the country of the Alamans (or Alemanni).[361] According to the narration of Peter the Deacon (1107–1159 A.D.), the librarian at Montecassino monastery and the author of *History of the Lombards*, the disease first affected the lymph nodes of the bubonic area and was accompanied by high fever, and patients died on the third day.[362] Another plague epidemic broke out in 571 and spread westwards to the Frankish Kingdom.[363] As Gregory of Tours (538–594 A.D.) mentioned, an open wound resembling a snake bite appeared in the buboes and armpits, and its poison killed the patient on the third day.[364]

Gregory of Tours also recounts the events of another bubonic plague epidemic that broke out in Rome in 590.[365] He draws his evidence from the testimony of his deacon, Agiulfus, who happened to be in Rome during the epidemic. According to him, the epidemic started in January 590 after the flooding of the River Tiber which had occurred two months earlier and destroyed many buildings. The epidemic quickly took on large dimensions and caused the death of countless residents, among them Pope Pelagius II (reign. 579–590), who was succeeded by Gregory the Great (reign. 590–604), a major figure of the Roman Catholic Church. The new Pope called on all the people of Rome to participate in a great litany; seven different processions would end up in the Basilica of Virgin Mary. During these processions, 80 people collapsed and died, but the religious zeal prevailed over the fear of plague, and all participants continued chanting until they reached the meeting point. The spread of this epidemic is possibly related to that of the Exarchate of Ravenna. Paul the Deacon points out that the epidemic of Ravenna was similar to the epidemic that occurred 30 years earlier, apparently implying the epidemic of 565 (?) the epicenter of which had once again been Rome.

---

359 Stathakopoulos 2004: 293–294.
360 Stathakopoulos 2004: 150.
361 Paulus Diaconus, *Historia Langobardorum*, II.4 (eds Bethmann and Waitz 1878: 74).
362 Paulus Diaconus, *Historia Langobardorum*, II.4 (eds Bethmann and Waitz 1878: 74).
363 Gregorius Turonensis, *Historiarum Libri*, IV.31 (A. D. 571) (eds Krusch and Levison 1937: 165.12–166.8).
364 Gregorius Turonensis, *Historiarum Libri*, IV.31 (A. D. 571) (eds Krusch and Levison 1937: 165.12–166.8).
365 Gregorius Turonensis, *Historiarum Libri*, IX.23 (A. D. 590) (eds Krusch and Levison 1937: 515).

We have already mentioned that, during the campaign of the Avars in Thrace in 598, an epidemic had broken out in Drizipera and affected Constantinople and Bithynia in the following year. This epidemic is believed to have spread throughout Asia Minor, whereas the report of an epidemic in Syria in 599 may be related to the outbreak of the disease on the East Mediterranean coast. Pope Gregory's correspondence with Venantius, a patrician of Palermo, in 601–604, and the Bishop of Carthage Dominicus in 591–601, reveals the existence of a plague epidemic in Italy and North Africa, which could also be related to the aforementioned epidemic on the East Mediterranean coast in 599.[366]

During the seventh century, Rome was struck by epidemics between 608–615, 664–665, and 676–678, while Pavia was affected in 680.[367] It is worth noting that the most famous victim of the epidemic of 664–665 was Wighard, the Archbishop of Canterbury (d. 660), and his entourage, who had traveled to Rome in order to be ordained an archbishop by the Pope. The last report of bubonic plague in Sicily, Calabria, Naples, and perhaps Rome dates to 745–746, and may be related to the report of Theophanes regarding the emergence and spread of the disease from the West to the Greek islands.

Middle East and Africa

The Middle East was a battlefield during the Persian Wars won by the Emperor Justinian. Given that both the Byzantines and Persians breached the peace agreements made from time to time, the war continued until the eventual peace agreement of 591. In the early seventh century, the Persians launched a major campaign and, between 611 and 620, they managed to seize Syria, Palestine, Egypt, and Libya from the Byzantines.[368] A great figure in the history of Byzantium, the Emperor Heraclius, launched a counterattack to recover the lost provinces (622–630). The decimation of the Avaro-Persian army in front of the walls of Constantinople in 626, and the overwhelming defeat of the Persians in Nineveh in 627 are considered as two events of paramount importance. The recapture of Syria, Palestine, Egypt and Libya by the Byzantines was completed in 630. But the exhausted Byzantine forces were unable to contain the new power that sprang from the deserts of Arabia. Two years (634–636) sufficed for the Arabs to conquer Syria and Palestine. The cultural and religious history of the Byzantine east was sealed forever on August 20, 636, after the incredible destruction of the imperial army in the battle of Yarmuk.[369] After this defeat, almost

366 Gregorius Papa, *Registrum Epistularum*, 2.3 (ed. Norberg 1982: IX.232, and X.20); Wispelwey 2008: 313, 1118.
367 Paulus Diaconus, *Historia Langobardorum*, VI.5 (eds Bethmann and Waitz 1878: 214.1–8); *Liber Pontificalis*, LXXXI (ed. Duchesne 1886:350.13–17).
368 Louth 2008: 226–228.
369 Louth 2008: 229–230.

all cities surrendered without a fight, and, by the year 642, the Arabs had conquered Egypt.

This change in the borders of the Empire, however, allows to expand the range of sources, since the epidemics are described in Syrian and Arabian texts that complement the Byzantine sources and fill temporal gaps. The term used in Syrian texts to describe epidemics is *mawtānā*, which is related to the Arab word *wabā* that translates into death or *thanatikon*. Syrian texts also use the word *shar'ūtā*, which is indirectly linked with plague as it denotes the swelling of some part of the body (bubonic area) and corresponds to the Arab *ta'ūn* that relates to the disease.[370] As in the case of Byzantine sources, however, *mawtānā* (*thanatikon*) is always equated to *shar'ūtā* (plague) in the Syrian texts, although the two cannot be fully verified. Accordingly, in the Arab texts, *wabā* (*thanatikon*) is used interchangeably with *ta'ūn* (plague).[371] The Arab chronographers tend to give epidemics the name of the ruler of a city or region.

The Byzantine wars in the Middle East coincide with the appearance of new epidemics. During the campaign of Heraclius, a plague epidemic characterized by a high mortality broke out in Persian Palestine.[372] In the same period, several Arab sources mention the plague epidemic of Shīrawayh in 628, which was named after the Persian King Shērōē (or Kavadh), son of the great adversary of Heraclius, King Chosroes II (reign. 590–628), who usurped the throne from his father and died from the disease in 629.

During the Arab attack in the Middle East (634–635) and following a strong earthquake, an epidemic of plague broke out and spread to many areas of Syria and Palestine.[373] A major bubonic plague epidemic exploded in the Arabian territories of Syria and Palestine in 639 and spread to Mesopotamia. Based on the sources, it is speculated that the epidemic had appeared in the previous year in Syria, wherefrom it spread using the Arab army as its vehicle, killing 25,000 Arab soldiers as well.[374] In 646–647 another epidemic broke out in the areas of Syria and Mesopotamia, as well as in the city of Marash (now Kahramanmaraş in Turkey) on the Arab-Byzantine border. The epidemic resulted in a terrible famine, and cases of cannibalism in the surrounding areas have even been mentioned. One more epidemic, originated in the city of Kufa on the banks of the Euphrates between 669 and 670, lasted for approximately eight months.[375] In 683, Syria was once again struck by the disease, while, in 687, plague spread from Mesopotamia to the Byzantine Asia

---

370 Morony 2007.
371 Conrad 1982: 268–307.
372 von Kremer 1880: 69–143; Michael Syrianus, *Chronica*, Liber XI, Cap. V (ed. Chabot 1899–1910: 4.414–415; Fr. transl. Chabot 1899–1910: 2.418–420, especially 419).
373 Michael Syrianus, *Chronica*, Liber XI, Cap. VIII (ed. Chabot 1899–1910: 4.421–423; Fr. transl. Chabot 1899–1910: 2.429–433, especially 431).
374 Elias Nisibenus, *Opus Chronologicum*, p. 133 (ed. Brooks 1910: 64.11–12; Conrad 1981: 51–93.
375 von Kremer 1880: 69–143.

Minor.[376] Egypt seems to have been affected by plague between 689 and 690, whereas the disease reappeared in Syria in 698. The following year it reached Mesopotamia. This plague became known in Arab history as the *Al-jārif plague* (*Torrential plague*). The name was euphemistic, as the Arab chronographers believed that the large number of deceased would suffice to fill the dry rivers so that they would look like torrents. According to the Arabic historian Abu'l-Hasan al-Madā'inī (752–843 A.D.), the disease that struck the city of Shawwāl wiped out its population, since 70,000 people died every day for three days. In his attempt to support his argument, the historian records the drama of two noblemen—Anas ibn Mālik and Al-Rahmān ibn Bakra—who lost 83 and 40 children, respectively.

It should be stressed, however, that the excessive numbers recorded by the Arabic historians, whose works are both religious and moralistic in nature, are usually sevenfold, thus giving their texts an eschatological and mystical character. Therefore, the mention of 70,000 deceased during the epidemic could be associated with the text of the Old Testament (2 Samuel 24: 13–15), where "the Lord sent a pestilence on Israel… and seventy thousand of the people died, from Dan to Beer-sheva" or with the verses of the *Qur'an* (Qur'an, Surah 69: *The Inevitable*) that describe the behavior of God towards the unfaithful: "Then into Hellfire drive him. Then into a chain whose length is seventy cubits insert him". Another typical description of the situation where the living did not suffice to bury the deceased stems from Syria in the year 704. In his universal chronicle, Michael the Syrian even suggests that the disease killed one third of the residents.[377] Two years later, plague reappeared in Syria, Kufa and Bosra. This plague became known as the *Plague of the Virgins*, because most of its victims were underage girls.

Between 714 and 715 a bubonic plague broke out in Egypt again. According to the sources, the deceased were more numerous than in the previous epidemic, probably referring to the epidemic of 689. It is believed that this epidemic may be related to the one that struck Syria the previous year.[378] Various Arabic sources refer to the epidemic of Adī ibn Artāh (718–719). It is speculated that this epidemic broke out when the Arab troops returned from the failed siege of Constantinople, wherefrom they transmitted the disease to Syria.

In 724, plague reappeared in Egypt, and it possibly spread to Syria and Mesopotamia in 725–726. The epidemic expanded quickly. The main symptom was the swollen lymph nodes.[379] In the period between 732–735, Egypt, Mesopotamia, Syria, and Iraq were once again affected by plague, although it is unknown whether this epi-

---

376 Elias Nisibenus, *Opus Chronologicum* 154 (ed. Brooks 1910: 74.16–17).

377 Michael Syrianus, *Chronica*, Liber XI, Cap. XVII (ed. Chabot 1899–1910: 449–452; Fr. trans;. Chabot 1899–1910: 477–483, especially 480).

378 Michael Syrianus, *Chronica*, Liber XI, Cap. III (ed. Chabot 1899–1910: 408–411; Fr. transl. Chabot 1899–1910: 408–413, especially 412); Severus, I.19. (ed. Evetts 1907: 273).

379 Elias Nisibenus, *Opus Chronologicum*, p. 164 (ed. Brooks 1910: 78.30–31); Theophanes, *Chronographia*, A.M. 6218 [= A.C. 725], P. 338 (ed. Classen 1839: 621.15–17).

demic is eventually related to Agapius' account of the events of the destruction of the Arab expeditionary forces in Asia Minor in 735.[380] The dating of the report on the epidemic of Egypt and other areas of North Africa believed to have taken place in 743–744 or 768 is problematic. According to the Arabic sources, it seems that plague appeared every year until 750, that is the year when the Umayyad Caliphate was abolished.[381]

The *Plague of Ghurāb* swap Syria and Mesopotamia in 744–745, and the Arabic historians speak of up to 20,000 deaths on a daily basis for a whole month.[382] These numbers certainly do not correspond to reality, but they obviously reveal the large extent of the epidemic. Finally, the *Plague of Salm ibn Qutayba* that broke out between 748 and 750, seems to have marked the end of the first pandemic in the Middle East. The epidemic originated in Mesopotamia and spread to Syria and Armenia, affecting Bosra from August 748 to August 749, with the highest mortality rates recorded between March and April 749. According to Abu'l-Hasan al-Madā'inī, (752–843) a major epidemic broke out in the city of Rajab during the Ramadan (March–April 749), and hundreds of people were buried every day. This plague was characterized as global by the Arabic historians, who recall that Syria and Palestine had been affected by plague since the ancient years before their conquest by the Arabs.[383]

The *Plague of Justinian* in the Kingdoms of Western Europe

The *Pars occidentalis* of the mighty Roman Empire followed its own historical course which was diametrically opposed to that of its Eastern counterpart. Throughout its borders it was invaded by various tribes. In 476, the Emperor Romulus Augustulus (reign. 475–476 A.D.) was dethroned by Odovacer (or Odoacer) (ca. 433–493 A.D.) in 476. This simply marked the official end of the Western Roman Empire and Romulus Augustulus became *ipso facto* its last ruler.[384] During the reign of Justinian, the Byzantines attempted to expand their dominion to the former territories of the Empire, fighting against the Ostrogothic Kingdom in Italy and against that of the Vandals in North Africa.[385] Further west, the Kingdom of the Visigoths on the Iberian Peninsula (412) and the Frankish Kingdom (480) on the territory of modern France had already been established. The Anglo-Saxons landed in Britain around the fifth

---

380 Michael Syrianus, *Chronica*, Liber XII, Cap. III (ed. Chabot 1899–1910: 4.482–484; Fr. transl. Chabot 1899–1910: 3.8–12, especially 8–9).
381 Severus, I, Cap. XVIII Michael I. A. D. 744–768 (ed. Evetts 1907: 97/[351]).
382 Michael Syrianus, *Chronica*, Liber XI, cap. XXII (ed. Chabot 1899–1910: 4.464–468; Fr. transl. Chabot 1899–1910: 2.505–511, especially 506 abd 508); *Chronica Zuqnīni, ad a. 1055–56* (ed. Harak 1999: 168–174).
383 *Chronica Zuqnīni, ad a. 1061–62* (ed. Harak 1999: 184–189).
384 Moorhead 2008: 196.
385 Moorhead 2008.

century, where they repelled the Celts towards Wales and Ireland, while the British North remained under the control of the native Scottish and Picts people.[386]

Spain: the Visigothic Kingdom

The *Chronicles of Zaragoza* are the only source of information concerning the appearance of bubonic plague on the Iberian Peninsula. Written by an unknown author, it constitutes an integral part of a larger chronicle, the *Chronicle of Victor of Tunnuna*.[387] The difficulty of their dating generates uncertainties regarding the chronology (exact year of the outbreak) and origin of the disease (whether plague was transmitted through the Eastern Mediterranean [542] or Italy [543]). The year 542 is considered more plausible due to the commercial relations between Spain and the Byzantine East. Plague arrived during the sailing season, with good weather conditions. As the main commercial stations of the Iberian Peninsula, Mediterranean cities like Barcelona, Valencia, Cartagena, and the cities along the Guadalquivir river were in direct contact with the large ports of Byzantium.[388]

The next outbreak of plague occurred between 580 and 582 or in 584, at a time when an epidemic was raging in the cities of Albi and Narbonne in Southern France.[389] It is also estimated that plague appeared in the Spanish ports in 588. Gregory of Tours considers the ports responsible for the spread of the disease to Marseille in the same year.[390] The advice of the Visigoth King Egica (reign. 687–701/703 A.D.) to the bishop of Narbonne not to travel to the Council of Toledo due to the outbreak of plague in the area serves as a source of information on the epidemic of 693. Two years before its conquest by the Arabs in 711, Spain was struck by plague epidemics between 707 and 711 (or between 709 and 711 according to other scholars).[391]

France: the Frankish Kingdom

According to Gregory of Tours, the disease of buboes (*lues inguinaria*) reached Arles, in the South, and quickly affected many other areas. This was in the year 543, when plague had already struck Italy, wherefrom it must have reached the Frankish Kingdom.[392]

---

386 Stenton 1971: 32–73.
387 Kulikowski 2007.
388 Rougé 1952.
389 Gregorius Turonensis, *Historiarum Libri*, IV.33 (A. D. 584) (eds Krusch and Levison 1937: 304).
390 Gregorius Turonensis, *Historiarum Libri*, IX.22 (A. D. 588) (eds Krusch and Levison 1937: 442).
391 Kulikowski 2007: 150–170.
392 Gregorius Turonensis, *Miracula*, VI *De Sancto Gallo Episcopo*, 6 (A. D. 543) (ed. Krusch 1885/1969: 234.15–16).

In 571, plague broke out at the heart of the Frankish Kingdom, in central and southeastern France, in the cities of Lyon, Chalon, Dijon, Clermont, and Bourges. Bourges had allegedly already been struck in 563.[393] The wide and rapid spread of this epidemic was apparently due to the dense river network of central France.[394]

We have seen that plague appeared in the major port of Marseille in 588. According to Gregory of Tours, the Spanish merchants were responsible for its transmission. The residents of Vivier and Avignon were severely tested by the epidemic of 590 (*lues inguinaria*), the year when the great epidemic of Rome described by deacon Agiulfus to Gregory of Tours broke out.[395] In the following year, plague reappeared in Marseille; it was apparently related to the epidemic of Avignon ("At in Galliis Masiliensim provintiam morbus saepe nominatus invasit", "In Gaul, the often mentioned plague struck the province of Marseille").[396] Marseille and the broader region of Arles were simultaneously affected in 599–600, 630, and 654–655.[397] The epidemic of Narbonne in 694 signaled the end of the first pandemic in the Frankish Kingdom.[398]

The Anglo-Saxon Kingdoms and Ireland

In the second half of the seventh century, the English kingdoms as well as the whole of the British Isles were hit by major plague epidemics. Plague appeared in Ireland for the first time in 544. It is speculated that it spread there through the trade routes with the Frankish Kingdom during the epidemic of 543. Gildas (c.482-c.570), the British scholar and author of the history of Britons *De Excidio et Conquestu Britanniae*, attributes the great epidemic to the sins of the British people.[399] In the Irish and Welsh chronicles, various terms are used to describe the disease, such as *pestis flava*, *lues flava*, *y fad felen* or *lallwelen*, and mainly the Celtic word *blefed* and its variations *belfeth* and *belefeth*.[400] The etymology of the latter term is unknown, although it is believed that the Celtic prefixes 'ble-', 'bel-' and 'bele-' denote the yellow color.[401] Considering that the Latin term *pestis flava* (*yellow plague*) is also used in the same texts, the identification of the disease is deemed problematic. A controversial source dating from 547 A.D. informs us of a 'great death' that occurred in Wales. An epidemic broke out in the central part of Ireland (in the area of Shannon) in 550,

---

393 Le Goff 1969.
394 Le Goff 1969.
395 Gregorius Turonensis, *Historiarum Libri*, X.23 (A. D. 590) (eds Krusch and Levison 1937: 515).
396 Gregorius Turonensis, *Historiarum Libri*, X.25 (A. D. 591) (eds. Krusch and Levison 1937: 517).
397 Le Goff 1969.
398 Le Goff 1969.
399 Dooley 2007.
400 Dooley 2007.
401 Dooley 2007.

which deserted the monasteries of the region.[402] The most representative scholar of the Anglo-Saxons is Beda Venerabilis (672–735 A.D.) who, in his work *Historia Eccle-siastica*, describes in a dramatic manner the first epidemics of Britain in 642 and 664, adding that the epidemics had killed thousands of people at the same time in Hibernia (Ireland).[403] His testimony is important because Beda presents the Anglo-Saxon point of view of the epidemic of Rome (664), where the Archbishop of Canterbury Wighard lost his life in 660. This epidemic is also interesting because of the restoration of pagan beliefs it generated, since many people saw the disease as the result of the wrath of the old gods due to the mass Christianization. After the end of the great epidemic of 664, fragmented reports imply sporadic outbreaks of plague in 666, 672, 675 and 679–680.[404]

The epidemic of 679 in Ireland was characterized by the high mortality of children, prompting the writers to describe it as the *Plague of Children* or the *Third Great Plague*.[405] The *Plague of Children* reappeared in Ireland in 682–683, and the last major bubonic plague epidemic in Britain broke out between 684 and 687 in the kingdoms of Essex, Northumbria, and Mercia.[406]

For many decades, the studies of the different aspects of the "Europization" of the Justinian's plague had as a common starting point a biological event with a catastrophic impact. We need to mention here that the real historical consequences of the Justinian's plague upon the Mediterranean region and the European continent in economic and demographic terms is nowadays questioned. The old concept that Justinian's plague was the event that triggered great historical changes in Late Antiquity and a great catastrophe for Europe is now revised on the basis of a renewed evaluation of the historical sources. Current research claims indeed that the maximalist view of the direct and indirect primary sources and archeological findings led to an overestimation of the real scale and impact of the plague in terms of institutional and cultural changes.[407] Moreover, the quantitative analysis of the original narratives, the promulgation of laws, the issuing of coinage, land cultivation, and other proxy data cast doubts on the demographic decline. In a recent study, the historian Lee Mordechai of the Hebrew University in Jerusalem, suggested that the high mortality of urban centers like Constantinople or Rome was the exception, and not the rule. A possible explanation of the overestimation of the decline of European popu-

402 Dooley 2007.
403 Beda, *Historia ecclesiastica*, Liber III, Caput XIII, § 186 [A. D. 642], and Caput XXVII, § 240 [A. D. 664] (ed. Stevenson 1838: 185, 231), with mention of the disease in the title III.XIII ("Ut in Hibernia sit quidam per reliquas ejus a mortis articulo revocatus") and III.XXVII ("Ut Egberct vir sanctus de natione Anglorum, monachicam in Hibernia vitam duxerit").
404 Dooley 2007.
405 Russell 1976.
406 Maddicott 2007.
407 Mordechai 2019a.

lation could be the tendency to use the *Black Death* as a model of mortality in the case of Justinian's plague.[408]

408 Mordechai 2019b.

Chapter 6
The Second Plague Pandemic

Europe Taken by Surprise

According to existing data, the second plague pandemic started in the Crimean peninsula in 1346 and—more specifically—in the Genoese colony of Kaffa.[409] Kaffa (now Feodosia in Crimean peninsula) was the former Byzantine Theodosia, which became part of Genoa in 1266. In 1340, the city was a major commercial hub, and the Genoese controlled the trade routes of both the Black Sea and Central Russia through the Don River. The city was crowded and fortified with double walls, enclosing 6,000 houses inside the inner wall, and another 11,000 within the outer wall.[410] Kaffa was a really impressive and cosmopolitan urban centre, the population of which consisted of Genoese, Venetians, Greeks, Armenians, Turks, Jews, and Mongols. The traditionally tense trade relations between Genoa and the *Golden Horde* of the Mongols culminated in the siege of Kaffa, which started in 1344. The loss of 15,000 of his men forced Khan Jani Beg (d. 1357 A.D.) to withdraw in order to regroup. The siege was resumed the following year. This time, however, both besiegers and besieged had to face something totally unexpected. While the siege entered its third year, the momentum of the Mongols suddenly declined, as hundreds of warriors were dying every day due to a strange illness. The Italian chronicler Gabriele de' Mussi (c. 1280 – 1356) recounts that "the Tartars, amazed and upset by the magnitude of the desolation and illness and having realized that they had no hope of salvation, lost all interest in the siege. But they ordered to chop up the dead and throw them in the city, in order to make the Christians die from the stench".[411] The residents would throw the shredded bodies back in the sea, but soon the first deaths began to occur in the city. The patients suffered from "swollen glands in the armpits and groin by coagulated juices, followed by fever with chills" that resulted in quick death. [412]

Upon lifting the siege of Kaffa, the Genoese galleys were once again free to sail. Medieval legends speak of three, four, nine, or twelve galleys that set off on the fatal journey, but the exact number of ships that left Kaffa after the siege remains unknown. The first station and—at the same time—the first victim was Constantinople and the Genoese colony of Peran.[413] In October 1347, the disease reached Sicily (Mes-

---

409 McNeill 1976: 132–176.
410 Wheelis 2002.
411 Derbes 1966.
412 Derbes 1966.
413 Kelly 2005: 79–101.

https://doi.org/10.1515/9783110613636-009

sina and Catania) where—despite the sanctification of water and prayers to the holy relics—it began to wreak death indiscriminately.[414]

According to the Florentine chronicler Giovani Villani (ca. 1276/1280 – 1348), "Having grown in strength in Turkey and Greece and having spread thence over the whole Levant and Mesopotamia and Syria and Chaldea and Cyprus and Rhodes and all the islands of the Greek Archipelago, the said pestilence leaped to Sicily, Sardinia and Corsica and Elba and from there soon reached all the shores of the mainland. And many lands and cities were made desolate" [415].

In late November, the disease appeared in Marseille, reaching Genoa and Venice in January 1348.[416] As a result, the mainland of Italy and France were overtaken by the disease between January and August 1348. Lawlessness and anarchy prevailed.[417] In Siena, an urban hecatomb led to the death of 52,000 people.[418] In Paris, the Royal College of Medicine published an essay entitled *Compendium de epidemia per Collegium Facultatis Medicorum Parisius Ordinatum*, in which the French physicians affirmed that the arrival of plague should not have surprised anyone, since three years earlier, "at one o'clock after midday on the twentieth of March, 1345, Cronus, Zeus and Mars aligned themselves in the House of Aquarius, creating poisonous gases that slowly descended to Earth...".[419]

While England and France were entangled in the Hundred Years' War, the disease reached Bordeaux in August 1348.[420] At that time, Princess Joanna (1333/4 – 1348), daughter of King Edward III of England (reign. 1327– 1377), accompanied by her brother, the legendary 'Black Prince' Edward (1330 – 1376), was in Bordeaux en route to Spain, where she was bound to marry the successor to the throne of Castile, Pedro (reign. 1350 – 1369). The fifteen-year-old princess died from plague in Bordeaux before the marriage that could possibly have altered the military balance in the war against France. King Edward was soon faced with an enemy far deadlier than the French, as plague crossed the Channel in the summer of 1348 and it spread throughout England by spring 1349.[421] In 1349 London, "the shops are closed, very few people walk on the streets and there is absolute silence throughout the city, besides some mourning voices and weeping of those who are burying their relatives".[422]

In September 1348, plague was in the areas surrounding Lake Geneva. It followed the Rhine and swept Southern Germany by the end of the year.[423] In May

---

414 Kelly 2005: 79 – 101.
415 Villani, *Nuova Cronica*, XIII, cap. LXXXIV "Di grante mortalità che ffu in Firenze, ma più grande altrove, come diremo apresso" (ed. Porta 1990 – 1991: 3.1576 – 1578); Carr 2015: 77– 78.
416 Kelly 2005: 79 – 101.
417 Gottfriend 1983: 33 – 54.
418 Bowsky 1964.
419 Herlihy 1997: 29 – 62.
420 Cantor 2001: 209 – 231.
421 Cantor 2001: 209 – 231.
422 Kelly 2005: 213.
423 Ziegler 2003: 65.

1349, riverboats carrying plague in their cargo spread the disease in the Netherlands through the Rhine canals.[424] In late 1349, all of Western and Central Europe were affected. In 1350, slowly but steadily, the disease started to move eastwards and only the regions of Bohemia and Poland were free of the disease.[425] The same year, the Kingdom of Norway was affected through an English ship anchoring in Bergen. A few months later, the neighboring Kingdom of Sweden was struck, while the disease reached Northeastern Russia and the Kingdom of Novgorod in 1351.[426] That first wave stopped in Moscow in 1352.[427] It is really impressive that the cyclical movements of the disease on its way to conquer the European continent did reach Moscow a whole five years later, although the city was located a mere 500 kilometers north of Kaffa.

When plague reappeared in the late Middle Ages, Europe was already in great turmoil. After defeating the Seljuk Emirs, the Ottoman Turks had settled in Asia Minor and were preparing to attack the Balkans, while the Iberian Peninsula was witnessing the Spanish *Reconquista* of Christian lands from Islam. England and France were on the verge of war, while the German Empire and the Pope were entangled in their traditional confrontation for the primacy of Europe's secular power. Italian cities were experiencing a civil war as the conflicts between the *Guelfi* (Pope's followers) and the *Ghibellini* (supporters of the German emperor) were almost a daily phenomenon. France and the Pope were trying to control the emergence of Hungary, which, in turn, was threatening the positions of Venice in the Adriatic. Germany and Poland were coveting Moravia and Bohemia, while the State of the Teutonic Order—as a genuine exponent of the German *Drang nach Osten* doctrine—organized Crusades against the Latvians, Lithuanians, Poles, and Russians. For the ordinary people, who were used to witnessing soldiers and mercenaries of all nationalities plundering their cities, the outbreak of plague was the final blow.

While medical astrology was attempting to identify the cause of plague, the disease was decimating one city after another. Data from specific cities must be mentioned to better understand the magnitude of the demographic catastrophe that took place in the West between 1348 and 1350: Bremen lost 70% of its population, Hamburg between 50 and 66%, Venice 60%, Genoa 58%, Perpignan 58%, Siena between 50 and 54%, Florence 52%, Marseille 50%, Exeter 49%, London 40%, Paris 30%, and Reims 25% of its population.[428]

The invisible *Y. pestis* caught Europe by surprise. The reactions of the inhabitants of medieval cities were rather hysterical. Fleeing a city was no more than a brief extension of life. The typical scenery of an infected city comprised death, flight, lawlessness and anarchy, a dissolute lifestyle and transient pleasures. The outbreak of

---

424 Cantor 2001: 67; Roosen 2019.
425 Cantor 2001: 73.
426 Langer 1975.
427 Langer 1975.
428 For these estimates, see Emery 1967; Ruffié 1984: 81–277; Cohn 2002; Olea 2005.

plague in Europe during the period 1347–1351 remains one of the greatest catastrophes to have tormented Humankind up to date. It is estimated that the epidemic caused the death of approximately 25 million people, that is, one quarter of Europe's population of that time.[429] In order to better comprehend the extent of the demographic catastrophe, it will suffice to say that, during World War II, the human toll amounted to 8% of the European population at the time.[430]

After 1351, the disease appeared periodically across Europe, terrorising people and leading them to decisions influenced by their ignorance and the superstitious beliefs of the time, such as mass persecutions against particular social or religious groups, or the participation in heretic movements of purification like the *Flagellati* (Flagellants).[431]

The Byzantine Empire in the Midst of the Two Pandemics

The history of the Byzantine Empire during the 600-years period between the last epidemic that struck Constantinople in 745 and the emergence of the *Black Death* in 1347 is adventurous and turbulent, with periods of both absolute power and omnipotence, and utter decline. By the end of the first pandemic, Byzantium had lost its possessions in Italy and the Middle East, and it was limited to Asia Minor and the Balkans. But, even in the Balkans, the Empire was threatened by the Serbs and Bulgarians, and its northern borders had moved to the axis Dyrrachium (Durrës in present-day Albania)–Scupi (Skopje, Republic of North Macedonia)–Serdica (Sofia, Bulgaria)–Mesembria (Nesebar, Bulgaria). The Byzantine diplomats eliminated the danger posed by the Longobards (Lombards) with the military help of the Franks and regained some of their possessions in Italy. Within the Empire, the Seventh Ecumenical Council, held in Nicaea in 787, ended the Iconoclastic Controversy. Social unrest had made the senior political and military officials aware of the need to save the Empire. The ninth century opened with the recovery of Byzantium, and the overthrow of the Empress Irene (reign. 797–802), who was replaced by Nikephoros I (reign. 802–811).[432]

The ninth and tenth centuries were characterized by the renaissance of education and Greek culture, which was dominated by the Patriarch Photios (810–891) and the missionary work of Cyril (826–869) and Methodius (815–885). The wars against the Arabs continued. The occupation of Crete by the Saracens in 828 was a major blow to the Byzantine dominance in the Southeastern Mediterranean. In 863, the victory against the Arabs in the Battle of Lalakaon (in current Turkey) marked the beginning of the offensive of Byzantium and of the decline of the Arab

---

429 Scott 2001: 87.
430 Bardet 1999: 13.
431 Ziegler 2003: 65–85.
432 Nistazopoulou-Pelekidi 1978.

power. In the North, the Christianization of Bulgarians gave the impression of the reconciliation of the two states, but it essentially served as a pretext for the 'Photian Schism' and the culmination of the conflict between the Papacy and the Patriarchate.[433]

The eschatological year of 1000 did not mark the end of the world as many believed at the time. On the contrary, the glory of the Empire increased under the guidance of Basil II (reign. 976–1025). By mid-eleventh century, the Empire expanded significantly after intense military campaigns in Asia Minor and the Balkans, by annexing areas of northern Syria, Mesopotamia, western Armenia, and Georgia. The Kingdom of Bulgaria was conquered and, once again, the Byzantine Empire reached the Danube after an absence of 400 years. In the West, despite the occupation of Sicily by the Arabs, the Byzantine administration was based in Calabria. The year 1054 marked the beginning of the Great Schism of the Eastern Orthodox and Roman Catholic Churches. Ever since the relations between the Pope and Byzantium have been extremely fragile. The account of the period from the death of Basil II until the ascension to the throne of Alexios I Komnenos (reign. 1081–1118) is rather daunting. Within a few decades, the military aristocracy that appointed emperors managed to discredit Byzantium. The dominance in Asia was lost forever, Byzantine possessions in Italy passed in the hands of the Normans, and the influence of the Empire in the Balkan Peninsula languished.[434] In 1099, the residents of Constantinople witnessed the stunning, yet bizarre spectacle of thousands of Crusaders heading to the Holy Land.

During the twelfth century, signs of collapse and state decay began to appear. In the Balkans, the alliance between Hungary and Serbia created new problems for the Empire, while the Emirates of the East were coveting Asia Minor.[435] In the middle of the century, an emperor with a catalytic presence on the European political scene tried to restore the grandeur of Byzantium. Influenced by the Crusades, Manuel I Komnenos (reign. 1143–1180) inaugurated a new and frugal image of the knight-king Byzantine ruler, which was free from the sometimes arrogant grandeur of his predecessors. On the basis of the doctrine of emperors' universality, Manuel used diplomacy and his relations with the German Empire to reverse the negative climate.[436] Despite his successes, the isolation of Byzantium at the end of the twelfth century was becoming increasingly obvious. The defeat of Byzantium in the epoch-making Battle of Myriokephalon (now in Turkey) in 1176 by the Turks, shook the edifice of Manuel to its foundations. The Empire lost its supremacy in the East, and it was definitively expelled from Italy, while it found itself in absolute isolation against a hostile alliance of Western powers.

---

433 Economidis 1978.
434 Stephenson 2008: 678–682, 686–690.
435 Korobeinikov 2008.
436 Tounta 2008: 29–137.

The thirteenth century was marked by the sack of Constantinople by the Crusaders in 1204 (Fourth Crusade) and the fragmentation of the Empire. The Latin kingdoms proved short-lived, but the negative impact of the occupation was bound to affect the future course of Byzantium, which reacted shortly afterwards. After the Latin conquest of 1204, the Byzantine ideal (including the reconquest of Constantinople) was preserved by three Greek states resulting from the fragmentation of the earlier unitary Empire: the Empire of Trebizond, the Empire of Nicaea, and the Despotate of Epirus.[437] Unfortunately, claims of hereditary rights in a virtually non-existent Byzantine Empire soon led Trebizond and Nicaea to a civil war. Eventually, the restoration was achieved by the reconquest of Constantinople in 1261 by Michael VIII Palaiologos (reign. 1261–1282) of the Empire of Nicaea.[438]

In an attempt to recover its former power, Byzantium moved into alliances with Genoa and Venice, and recognized the Papal primacy and the faith of the Catholic Church in the Council of Lyon in 1274.[439] With cunning diplomatic maneuvers and significant military victories, Michael managed to gradually extend his possessions in Greece, while the Byzantine fleet reestablished its dominance in the Aegean Sea. Michael's policy of unification soon created a staunch anti-unionist coalition, which undermined his undertakings and brought the Empire to the brink of civil war. After Charles of Anjou (King of Sicily 1266–1285) defeated the German Hohenstaufen imperial dynasty in south Italy, he became the defender of the Pope and the heir of the traditional ambitions of the Normans of Sicily, who aspired to conquer Constantinople.[440] While the powerful fleet of the Norman king was preparing to depart on the great campaign, an event of major historical significance took place that overturned the plans of the ambitious Charles. On the afternoon of 31st of March 1282, the bells of Palermo invited the faithful to the vespers, but, at the same time, this was the call to the great revolution of the Sicilians against the tyrant king Charles and the Normans, a revolution which had been organized with utmost secrecy by the Byzantine emperor and the king of Aragon. The fleet of Charles was set on fire, and the king himself barely managed to escape Palermo, whereas the incredible massacre of Norman soldiers became known in history as the 'Sicilian Vespers'.[441]

The restoration of Byzantium was rather short-lived, and Michael's successor, Andronikos II Palaiologos (reign. 1282–1328), was under permanent pressure by the enemies of the Empire. As the power of the Serbs and Ottomans increased dangerously, Byzantium was burdened by the irrational economic demands of the mercenaries who constituted the majority of the Byzantine army by then. The formerly powerful Byzantine golden coin (*hyperpyron*) was now made with a mixture of cheap metals and became undervalued, thus losing its appreciation on the markets

---

437 Miller 1908: 27–48.
438 Lyberopoulos 1999: 32–40.
439 Ostrogorsky 1968: 444–465.
440 Runciman 1958: 113–133.
441 Runciman 1958: 236–250.

and in the transactions with the West.[442] The shrinking and forfeiture of the formerly powerful Byzantine army which—according to the historian Nikephoros Gregoras (1295–1360)—provoked the laughter of its opponents, led the Empire to abjection, and the need to pay extravagant sums of money for the much-needed peace, thus plunging its economy even further. Andronikos II Palaiologos (reign. 1282–1328) sealed the condemnation of the Empire. In order to reduce public expenditure, he abolished the Byzantine navy and conceded the naval defense of Byzantium to the Genoese. One of the worst moments in the Byzantine history of the fourteenth century was the revolt of the Catalan mercenaries, who plundered Thrace, Mount Athos, Thessaly, and Central Greece before occupying the Frankish Duchy of Athens in a protest for the salaries they had not received.[443]

A few years before Constantinople was struck by plague, the situation of the Empire was rather pathetic, since the elder emperor Andronikos II Palaiologos was fighting against his grandson Andronikos, who further became the Emperor Andronikos III (reign. 1328–1341).[444] An ally of young Andronikos, Ioannis Kantakouzenos (1292–1383) would later describe—as a historian—the epidemic in Constantinople. In 1328, Andronikos III ascended to the throne, but Kantakouzenos was ruling the Empire behind the scenes. The two men were able to contain the Serbs and Bulgarians, and managed to implement some fiscal reform in the everyday life of the Empire. In 1341, Andronikos III died, and Kantakouzenos claimed the regency from empress Anna of Savoy (1306–1365), since the successor to the throne Ioannis V (b. 1332) was only nine years old.[445] In the ensuing years, a new civil war broke out between Kantakouzenos and empress Anna, and, by 1347, Kantakouzenos had risen to the throne of Constantinople. He treated empress Anna favorably, and became the spiritual father of young Ioannis, thus showing that he was the continuator of the Palaiologan dynasty.[446] But the price was too heavy and the historical consequences incalculable. During the civil war, Kantakouzenos allied with the Ottoman Turks and essentially owed them his crown. At that time, the Byzantine Empire encompassed the areas of Constantinople and Thrace, Monemvasia, Sparta, Mystras, Karytaina, and Argos in the Peloponnese, Thessaloniki, Serres, and Petritzos in Macedonia, as well as the North Aegean islands of Lemnos, Imvros, Tenedos, Samothrace, and Thasos.[447] Emperor Kantakouzenos felt unsafe with the Ottomans so close to him, and at the same time humiliated by the Genoese of Peran, who received 200,000 *hyperpyra* annually, that is, 87% of the revenue of the Bosporus customs.[448] These were the most serious problems that the new emperor was faced with, but none of them

---

442 Ostrogorsky 1968: 466–477.
443 Ostrogorsky 1968: 466–477.
444 Ostrogorsky 1968: 466–477.
445 Laiou 2008: 822–824.
446 Ostrogorsky 1968: 466–499.
447 Ostrogorsky 1968: 466–499.
448 Ostrogorsky 1968: 466–499.

mattered in the spring of 1347. The first galley from Kaffa was anchoring in the waters of the Bosporus, and soon Europe would plunge into chaos.

The *Black Death* in the Byzantine Empire (1347–1400)

The fourteenth century can be considered as the century of the mortuary ritual of the Byzantine Empire. Military defeats, dynastic controversies, civil wars, and natural disasters were further aggravated by the new ordeal of plague. Along with the other misfortunes of the time, plague exacerbated the sense of abandonment of the people, who considered themselves condemned by God. Despite their literary exaggerations, the accounts of the time are a valuable source of information concerning those epidemics. Unfortunately, however, the demographic data of the Byzantine historiography during the *Second Pandemic* cannot be compared to the plenitude of the sources in Western Europe. This greatly hinders the study of the epidemics of the time, since it is impossible to determine the dimensions of the catastrophe that took place.

Constantinople was the first city to be affected by plague in 1347, when the epidemic started to spread to Europe. The events of the epidemic are described by Ioannis Kantakouzenos, Demetrios Kydones (1324–1397), and Nikephoros Gregoras. When recounting the events of the period 1320–1356 in his *History*, Ioannis Kantakouzenos was no longer emperor, but had become a monk after his dethronement in 1354. In 1347, Demetrios Kydones had entered the service of Kantakouzenos and held his position until the abdication of the emperor. He was a firm proponent of an anti-Ottoman policy in combination with a military alliance with the kingdoms of Western Europe. Nikephoros Gregoras, whose work recounts the events of the period 1204–1359, was a protégé of emperor Andronikos II Palaiologos, whereas in the past he had served as an ambassador of Byzantium in the Serbian court.

The accuracy of Kantakouzenos' description was questioned in 1976 by the Austrian Byzantinist Herbert Hunger (1914–2000), who underlined the similarity of the narrative with that of Thucydides on the *Plague of Athens*, which had already been imitated by Procopius when he described the *Plague of Justinian*.[449] The same year (1976), however, the American Timothy Miller restored the rightness of the former emperor's testimony by showing that—although the literary style is indeed an imitation of Thucydides—the actual symptoms of the epidemic described refer to bubonic and pneumonic plague, which would have been impossible for Thucydides to know.[450]

The progression of the disease was fulminant and—according to Kantakouzenos —some patients died within the same day, while others died after a maximum of

---

449 Hunger 1976; Congourdeau 1999.
450 Miller 1976.

three days.[451] Similarly, Demetrios Kydones wrote that the population of the city was diminishing every day and the living were no longer enough to bury the dead.[452] All sources agree that the *thanatikon* or *disease of the buboes* affected all ages and social classes.[453] But, according to Kantakouzenos, when other diseases appeared, they had the same clinical picture and outcome, since "whatever the disease, it always led to the same results: (...εἰ δέ τις καὶ προέκαμνέ τι, πάντα εἰς ἐκεῖνο κατέληγε νόσημα...).[454]

Nikephoros Gregoras, as for him, noted[455]:

... οὐ μόνους γε μὴν ἀνθρώπους τὸ πάθος οὑτωσὶ διατελεῖ μαστίζον, ἀλλὰ καὶ εἴ τι κατ' οἶκον ὡς τὰ πολλὰ τοῖς ἀνθρώποις ἐκείνοις συνέζη τε καὶ συνῴκει τῶν ἄλλων ζώων. Κύνας φημὶ καὶ ἵππους καὶ ὀρνίθων παντοῖα γένη καὶ εἴ τινες ἐν τοῖς τῶν οἴκων τοίχοις οἰκοῦντες ἔτυχον μύες ...

... the disease does not affect humans only, but also the animals living alongside humans in their houses. Dogs, horses, all sorts of domestic birds as well as the mice that happen to live inside the house walls ...

The Byzantine sources describe two different aspects of the same disease. While in the sixth century Procopius stressed the delirium of patients, Kantakouzenos in the fourteenth century reported the coma, saying that "when it (the disease) reached the head, patients became speechless and were completely unaware of what was happening around them and fell into deep sleep" (καὶ ἐς κεφαλὴν τοῦ νοσήματος ἐμπίπτοντος, ἀφωνία κατείχοντο καὶ ἀναισθησία πρὸς τὰ πάντα τὰ γινόμενα καὶ ὥσπερ πρὸς ὕπνον κατεφέροντο βαθύν).[456]

But the account of Kantakouzenos points to the pneumonic form of the disease[457]:

... ἀλλ' εἰς τὸν πνεύμονα ἐμπίπτον τὸ κακὸν φλόγωσίς τε ἦν αὐτίκα πρὸς τὰ ἔνδον καὶ δριμείας ἐνεποίει ἀληγηδόνας περὶ τὰ στήθη, ὕφαιμόν τε πτύελον ἀνέπεμπον καὶ πνεῦμα ἄτοπον ἀπὸ τῶν ἔνδον καὶ δυσῶδες ...

... when the evil befell the lungs, there was immediately an internal burning and it [the disease] provoked a great pain over the chest and their sputum and breath from the inside was malodorous ...

Kantakouzenos describes a multifaceted disease. When the head is directly affected, the patient falls into lethargy and quick death follows. When it affects the lungs, the

---

451 Joannes Cantacuzenus, *Historiae*, IV.8 [A.C. 1347] P.730 (ed. Schopen 1832: 50.8–9).
452 Loenertz 1956: 122.
453 Bartsocas 1966.
454 Joannes Cantacuzenus, *Historiae*, IV.8 [A.C. 1347] P.730, 6–7 (ed. Schopen 1832: 50.6–7).
455 Nicephorus Gregoras, *Byzantina Historia*, XVI, A.M. 6855 [= A.C. 1347], P. 501 (ed Schopen 1829–1830: 2.798.2–6).
456 Joannes Cantacuzenus, *Historiae*, IV.8 [A.C. 1347] P.730 (ed. Schopen 1832: 50.11–13).
457 Joannes Cantacuzenus, *Historiae*, IV.8 [A.C. 1347] P.730 (ed. Schopen 1832: 50.16–21).

inflammation causes very strong chest pains, haemoptysis, and intense thirst. According to existing descriptions, abscesses appeared "everywhere in the upper and lower limbs and in more than a few cases on the jaws" (πανταχόθεν ἦν ἐπί τε ταῖς ὠλέναις ταῖς ἄνω καὶ κάτω, οὐκ ὀλίγοις δὲ καὶ πρὸς τὰς σιαγόνας). When ruptured, these abscesses released a malodorous black pus.[458] References to buboes on the tips of the thighs and arms coupled with haemoptysis can be found in the narrative of Gregoras too.[459] Kantakouzenos also refers to the appearance of skin lesions, as the appearance of black phlyctenae (pustules).[460]

In the same year, plague also struck Euboea, Lemnos, Thessaloniki, and Trebizond, in addition to the Venetian possessions in Euboea, Methone, and Koroni.[461] In 1348, the situation had not improved and the *unexpected rhomphaia of the thanatikon* struck Rhodes, Cyprus, and the inland of the Peloponnese.[462] Fourteen years later (that is, in 1361), plague reappeared in Constantinople, when the Ottomans occupied Didymóteicho (north-eastern Greece) and were preparing to conquer Adrianople (now Edirne in Turkey). The generalized degradation was evident from the large number of deceased, and the depressing impression that the epidemic was never going to stop was intense. Most of the nobles of Constantinople fled, and the city was abandoned to its fate. The epidemic continued to ravage the capital. In 1362 it reappeared in Pontus as a result of the commercial relations of the Empire with Trebizond. That same year, epidemics broke out in Crete, Cyprus, Lemnos, and the Peloponnese.[463] The epidemic cycle that started in 1361 continued until 1365, as the plague that struck Constantinople did not subside before 1364, whereas it affected the Peloponnese, Adrianople, and Crete until 1365.[464] In 1368, the *Great Plague of Ioannina* occurred which—according to the *Chronicle of Ioannina*—left many orphans and widows. The Serb Despot of Epirus Thomas Preljubović (reign. 1366–1384) forced the widows to get married with Serbian officials.[465]

The correspondence of Kydones with a friend of his reveals the existence of plague in Thessaloniki in 1372. In his reply—obviously overwhelmed by the news of the new epidemic—, Kydones recalls the appearance of the graves of the friends whom he lost in 1362, along with the death of his mother and two sisters. In that depressing letter, Kydones describes his residence as full of mourning for the deceased and the

---

458 Joannes Cantacuzenus, *Historiae*, IV.8 [A.C. 1347] P.731 (ed. Schopen 1832: 51.1–3).

459 Nicephorus Gregoras, *Byzantina Historia*, XVI, A.M. 6855 [= A.C. 1347], P. 501 (ed. Schopen 1829–1830: 2.798.12–14 with the mention of buboes).

460 Joannes Cantacuzenus, *Historiae*, IV.8 [A.C. 1347] P.731 (ed. Schopen 1832: 51.4–5).

461 Thiriet 1958: 63–64.

462 Thiriet 1958: 214–216.

463 Loenertz 1956: 145–149; Thiriet 1966: 253; *Chronica Byzantina Breviora*, 89.I.2 (ed. Schreiner 1975: 1.619).

464 *Chronica Breviora*, 15 (ed. Lampros 1932: 31.12–13); *Chronica Byzantina Breviora*, 7.II.14 (ed. Schreiner 1975: 1.66).

465 *Chronica Ioanninensis*, 12 (ed. Vranousis 1962: 82–84).

fear of not knowing whether one would remain alive until the evening.[466] The next reference in 1374 is again associated with the Despotate of Epirus, which witnessed the *Great Plague of Arta*, while the epidemic spread to the Peloponnese.[467]

In 1375, the epidemic went beyond the boundaries of Arta and reached Ioannina. The city paid a high price for its overpopulation and impoverishment resulting from the misguided and authoritarian policy of Preljubović. It is difficult to estimate the population of Ioannina at the time of the epidemic. Based on the archaeological and topographical data, it is estimated that the city had at least 1,000 residences and a population of 5,000 inhabitants.[468] Ioannina was one of the best fortified cities in Greece, as the lake around the castle made it virtually impenetrable. But its restricted size coupled with the massive influx of villagers who sought protection in the castle from the Albanian raids, exacerbated the residential problem of Ioannina.[469] In that same year, the Albanian ruler of Arta, Pjeter Ljosha (reign. 1359 – 1374), died from plague, and the ruler of Achelous, Gjin Bue Spata (d. 1399), seized the opportunity to take control of the city and unite the two Despotates. This new Albanian Despotate became a permanent threat to the Serbian Despotate of Ioannina, and the city of Ioannina was constantly besieged, even during plague epidemics.[470] At the same time, plague appeared in Crete and tormented its residents until 1376, whereas Mount Athos was affected in 1378.[471] Apparently, in 1379 – 1380 a plague epidemic broke out in Galatas and possibly in the Genoese colony of Peran.[472] Six years after the last epidemic, the Peloponnese once again experienced a plague outbreak in 1381 – 1382.[473]

A series of letters by Kydones indicate that Constantinople witnessed one more epidemic in 1386. While the capital appeared to be cut off from the rest of the world, Kydones complained because he was unable to reach Italy, where he would negotiate with the West to seek its assistance against the Ottomans, although it is hard to establish whether that inability was due to a ban on leaving the port or to the lack of seamen to man the ships.[474] It seems that Athens was also affected by an epidemic in 1388. Crete witnessed one more outbreak in the following year.[475] In 1390, an epidemic broke out in the Peloponnese.[476]

---

466 Loenertz 1956: 144.
467 *Chronica Breviora*, 19 and 27 (ed. Lampros 1932: 36.10 – 14, and 46.15, 17 – 18, 20, 24); *Chronica Ioanninensis*, 15 (ed. Vranousis 1962: 85).
468 Kordosis 2003: 25 – 49.
469 Kordosis 2003: 25 – 49.
470 Kordosis 2003: 25 – 49.
471 *Acta Koutloumousiou*, 38 "Troisième testament de Chariton" (dated 1378) (ed. Lemerle 1946/1988: 135).
472 Loenertz 1956: 122; Lyberopoulos 1999: 32 – 40.
473 Loenertz 1956: 107.
474 Idem: 263,273,285,317.
475 Setton 1975: 147 – 148,181 – 183.
476 *Chronica Byzantina Breviora*, 33.I.15 (ed. Schreiner 1975: 1.244).

The year 1391 was harsh for Constantinople as it suffered many disasters: emperor Ioannis V Palaiologos died in February, the city was hit by an earthquake on the holy day of the Dormition of the Mother of God (August 15), in the autumn the Turks besieged Constantinople, and throughout the winter the city was tormented by plague.[477] An epidemic in Cyprus in 1393 is documented, and Venetian sources mention an epidemic in Candia (modern Heraklion, Crete) that completely decimated the Venetian Guard.[478] At the turn of the century, the disease reappeared in the Peloponnese, Koroni, and Methone in 1399.[479]

The *Black Death* in the Byzantine World of the 15th Century

In the early fifteenth century, the Empire shrunk even further. Its borders extended to Rhaedestus (now Tekirdağ in Turkey), a few dozen kilometers westward from Constantinople, while the flag of the Palaiologan Dynasty was now waving only in the islands of Imvros (now Gökçeada), Tenedos, Samothrace, Thasos, and the castle-town of Mystras in the Peloponnese.

Following the outbreak of 1399, plague continued to afflict Koroni in the Peloponnese, and a Venetian document suggests that the epidemic broke out in Corfu in 1400.[480] Given that the Venetian galleys were moving freely in the Ionian Sea and anchored in the naval bases of Koroni and Methone among others, it should not be considered unlikely that the disease was transferred from the former to the latter in 1402.[481] In the early fifteenth century, the collapsing Empire was offered a brief extension of life as the Mongol troops of Tamerlane (1336–1405) crushed the Ottomans of sultan Bayezid (1360–1403) at the battle of Ankara in 1402, thus giving the Byzantines the opportunity to reorganize. Our sources are not clear as to the existence of plague, but Ruy González de Clavijo (d. 1412), the head of the Spanish embassy to the court of Tamerlane, points out in his report that he spent some time in Chios in 1403, because of a large outbreak of an infectious disease in Gallipoli.[482]

In 1408, an extended epidemic wave seems to have started, which lasted until 1413. In 1409 it was in Cyprus and Constantinople. The disease was first located in the island of Karpathos and, close to it, in Crete.[483] There is a report on plague in Cyprus for the following year as well, and the disease once again reached Koroni and

477 *Chronica Breviora*, 19 (ed. Lampros 1932: 36.10–14); *Chronica Byzantina Breviora* 33.I.15 (ed. Schreiner 1975: 1.244).

478 Leontius Machairas, *Chronaca*, § 137 (ed. Sathas1873: 113.5–6); Noiret 1892: 92–96.

479 Noiret 1892: 101; *Chronica Breviora*, 27 (ed. Lampros 1932: 47.36, 40, 45, 47–48); *Chronica Byzantina Breviora*, 33.I.21 (ed. Schreiner 1975: 1.245).

480 Sathas 1880-1890: 3.39.

481 Asonitis 1987: 11.

482 Clavijo 1859: 26.

483 *Chronica Byzantina Breviora*, 9.II.41, and 33.I.25 (ed. Schreiner 1975: 1.97 and 1.246).

Methone wherefrom it spread to Corfu.[484] During the service of the Venetian *Bailo* of Corfu, Michele Malipiero (1409–1410 A.D.), sent a request addressed to the Venetian Senate in 1410 raises suspicions of several victims in the ranks of the guard of Corfu. According to that request, Venice was asked to send an urgent expedition of archers (*balistari*) in order to strengthen the guard after its losses during the last epidemic.[485] Plague seems to have continued to scourge Crete, Koroni, and Methone until 1411, whereas it seems to have lost its impetus in Corfu around 1413.[486] It is estimated that Cephalonia was also hit by a plague epidemic in 1416.[487]

In 1417, the authorities of Constantinople decided to stop the operation of the port due to the extent of the epidemic, which seems to have spread to the coast of the Black Sea as well.[488] In 1418 and 1419, the disease spread to Crete, the Peloponnese and Epirus, while, in 1419, it reappeared in Cyprus.[489] Plague continued to torment the residents of Corfu and Cyprus until 1420.[490]

A new wave of plague started in Constantinople in 1421–1422, and spread to Cyprus, the Peloponnese, and possibly Thessaloniki.[491] The siege of Constantinople by the Ottomans in 1422 coincided with the plague epidemic. Despite their desperate situation, the defendants were able to withstand the raids until the exhausted besiegers were forced to retreat. That wave of plague completed its course in 1423, reaching Boeotia, Athens, and Megara, and it is believed that, in the same year, an epidemic broke out for some time in Thessaloniki as well.[492] In 1426, Negroponte (Chalkida) was added to the long list of cities struck by plague.[493]

The Peloponnese, more specifically Patras, was once again tormented by plague in 1431, while the disease appeared in Constantinople, too.[494] The most severe outbreak, however, occurred in Constantinople and Trebizond in 1435.[495] In 1435–1436, the negotiations between the Greeks and Latins regarding the organization of the upcoming Council of Ferrara for the union of the two Churches were marked by yet another outbreak of plague. Plague appeared during the Council itself a few years later. Among the numerous deceased was the papal envoy Simon Fréron, who had traveled to Constantinople for the preparations of the Council. The other

---

484 *Chronicon Breve*, A.M. 6918 [= 1410] (ed. Bekker 1834: 517.9–10).
485 Sygkellou 2008: 103–106.
486 Thiriet 1959: 129.
487 Sathas 1880-1890: 3.140.
488 *Chronicon Breve*, A.M. 6926 [= 1418] (ed. Bekker 1834: 517.15).
489 *Chronica Breviora II*, 27.36–48 (ed. Lampros 1932: 47); Machairas, *Chronicle of Cyprus*, § 644 (ed. Sathas 1873: 381.23–24); Noiret 1892: 92–96.
490 Leontius Machairas, *Chronaca*, § 644 (ed. Sathas1873: 381.23–24); Colin 1981: 98–99.
491 Leontius Machairas, *Chronaca*, § 648 (ed. Sathas 1873: 382.1 ab imo-383.1); *Chronica Breviora*, 32.2–4 (ed. Lampros 1932: 61); Georgius Sphrantzes, *Chronica*, VIII.1 (ed. Maisano 1990: 18).
492 Buchon 1845: 271–272, 280–281.
493 Sathas 1880-1890: 3.313–314.
494 Georgius Sphrantzes, *Chronica*, XXI.8 (ed. Maisano 1990: 72).
495 Lampros 1924: 284–285, 289; Thiriet 1961: 50.

envoy of the Pope, John of Ragusa (d. 1443), fled to the countryside in order to save himself from the epidemic that ravaged the city, while Venice prohibited its ships from approaching the coasts of Constantinople due to plague.[496] In 1438 an epidemic broke out in the suburbs of the capital, and an epidemic was reported in Nicosia.[497] The last reference to plague in Byzantine Constantinople and its suburbs dates to 1445, when an epidemic was also reported in Chios.[498] In 1448, the disease must have been present in the city of Silivri in Thrace, the Peloponnese, and Chalkida.[499] The last epidemic of plague in Greece occurred in Corfu before the fall of Constantinople; it was recorded by Venetian sources in 1450.[500]

The lack of accurate demographical data coupled with literary excesses raise questions as to the actual mortality rate of an epidemic. The references to epidemics are often indirect and do not include any description of their extent or mortality rate. The proposals and assumptions resulting from the study of a broad range of sources only offer a general picture of the tendencies in the fluctuation of populations. To some extent, the regulations and decisions found in Venetian state documents and notarial acts offer a general picture of the economic and demographic disaster that occurred in some cities. Indicative examples are the special arrangements of debts to the state in 1348 as a motive for residents to return to the ruined city of Rethymno in Crete, the tax exemptions for the residents of the same city (exemption from installation fees to the newcomers in the city) in 1376, or the measures of the same year concerning the possibility of compromise and debt repayment in installments for the inhabitants of Candia.[501] Another example is that of Methone and Koroni in the Peloponnese, which were repeatedly struck by plague. These two station-cities were of key importance to Venice. The demographic catastrophe caused by plague forced the authorities to send a new population of colonists there, as the Venetian regulations of *Statutes and Capitulations* (*Statuta Capitula*) reveal.[502]

Regarding the epidemic of 1347, both Kantakouzenos and Gregoras are content with the vague information that many people died of this disease in Byzantium (then mostly reduced to the city of Constantinople).[503] The estimates of losses in Constantinople in the order of two thirds of the population are probably exaggerated. In his account of the epidemic of Cyprus in 1348, the Cypriot Leontios Machairas (1360/

---

496 Congourdeau 1993: 21–41.
497 Leontius Machairas, *Chronaca*, § 707 (ed. Sathas 1873: 408.15–17).
498 Paidousis 19396.
499 Laonicus Chalcocondyla, *Historiae*, De rebus Turcicis VII, P. 180/V. 141 (ed. Bekker 1843: 341.5–10); Thiriet 1961: 50; Georgius Sphrantzes, *Chronica*, XXVI (ed. Maisano 1990: 98).
500 Asonitis 1987: 14.
501 Kostis 1995: 221–222. For more on the economic response of medieval states, see Cohn 2007.
502 Miller 1908: 300.
503 Nicephorus Gregoras, *Byzantina Historia*, XVI, A.M. 6855 [= A.C. 1347], P. 501 (ed. Schopen 1829–1830: 2.798.1–5); Joannes Cantacuzenus, *Historiae*, IV.8 [A.C. 1347] P.731 (ed. Schopen 1832: 51.9–10).

80—after 1432) estimates that half of the island's population died, but, in his report of the 1393 epidemic, he just makes the general statement that many people died.[504]

The sources contribute to the understanding of the disruption brought about by the epidemic. In 1361, Kydones described the flight of the nobility and the abstention of the clergy from its duties, as the priests refused to perform burials and chant their prayers to the dead. It is only reasonable that the psychological impact was no less severe. In 1348, Kantakouzenos pointed out that, on the one hand, the epidemic led some people to repent and to distribute their goods to the poor, whereas, on the other hand, others were led to absolute despair or utter impunity.

Although physicians had long argued that plague could not be transmitted from one patient to the other, but that it rather affected all the people in an area, as all were exposed to the same infectious agent (that is, the air of the region), the ordinary people preferred to believe what they could see with their own eyes. In 1347–1348, the young Kydones wrote that the people avoided each other out of fear of transmission of the disease.[505] Kydones was not interested in a scientific explanation, but rather focused on the moral degradation that ensued, namely that a father did not bury his children, but simply abandoned them because of the fear of transmission, and *vice versa*. During the same epidemic, Gregoras exaggerated by saying that nobody helped anybody any more. Observing the evolution of the epidemic of 1347, Kantakouzenos suggested that the contact with patients was a major cause of proliferation of plague.[506] In so doing, he probably reflected the general perception of the residents of Constantinople.

Physicians' helplessness in the face of major epidemics has been a fact ever since the time of Thucydides' time. What the Byzantine scholars owe to Procopius is perhaps his obsession with the absurdity of the calamity, which confuses medical reasoning, as Kantakouzenos argues.[507] Kydones is more poignant. On many occasions he criticizes the medical community of his time for the impotence and indolence of physicians during the epidemic of 1347. Many physicians dropped out and the health institutions were marooned, while those who stayed admitted that they were unable to help, and they died in desperation and shame, or devised unsound excuses to make up for their impotence. By 1361, the opinion of Kydones had remained the same; he suggested that the patients were more afraid of the physicians rather than of the disease itself, because all they did was to defame and blame each other. Their therapies were useless and only caused further damage.[508] Perhaps this criticism is due to the bitterness of a man who had seen his mother and two of his sisters succumb to the epidemic within a few months (the former two in Thessaloniki

---

504 Leontius Machairas, *Chronaca*, § 622 (ed. Sathas 1873: 374.5–6).
505 Loenertz 1956: 122.
506 Joannes Cantacuzenus, *Historiae*, IV.8 [A.C. 1347] P.731 (ed. Schopen 1832: 51.18–20).
507 Joannes Cantacuzenus, *Historiae*, IV.8 [A.C. 1347] P.732 (ed. Schopen 1832: 52.5–10).
508 Demetrius Cydones, *Epistulae*, 25 and 31 (ed. Boissonade 1844: 304–306 [esp. 305.22–25], and 315–318 [esp. 317.28–31]).

and the latter in Constantinople). In fact, Kydones criticized the physicians mainly because they were trying to discover natural causes in what he perceived as an act of God. Nevertheless, the Byzantine medical community tried its best to confront plague, even in an unorthodox manner. Under no circumstances was the etiological and therapeutic approach of the Byzantine physicians limited by theories of astrological effects or Divine wrath. For instance, the pharmacist Nicholas Myrepsos of uncertain epoch (13th or 14th century ?) proposed a specific treatment for plague based on a patch impregnated with nitro, cumin and rosin. A more complex formula was based on litharge, myrrh gum, bdellium, mastic, and aristolochia, which relieved patients from the pain of swollen lymph nodes.[509]

Both the letters of Kydones—written at the time when the events took place— and the account of Kantakouzenos—albeit written later—reveal the psychological turmoil of the humans who found themselves in the heart of the epidemic. Kydones, in one of his letters to one of his friend who was a monk, said that his soul was in fear and noted the horror of the death. In 1361, Kydones—faced with the same misfortune— compared Constantinople with a massive tomb, and declared that the residents had lost all hope of seeing the destruction cease before completing its work and exterminating the entire population. A few months later, he listed all the friends he lost, along with his mother and sisters. In the same letter, he confessed that he felt dead himself, suffering from headaches, tachycardia, shortness of breath and insomnia, and feared that he was losing his mind—in other words, he exhibited all the symptoms of anxiety that reflexively affected all of the city's inhabitants. Maybe that fear made sense after all, since plague is not selective, unlike other infectious diseases, but affects all ages, sexes and social strata, as Gregoras put it: "… the disease affected men and women, rich and poor, old and young" (… ἐπενέμετο δὲ τὸ πάθος ἐπίσης ἄνδρας καὶ γυναῖκας, πλουσίους καὶ πένητας, γηραιοὺς καὶ νέους).[510]

With the fall of Constantinople and the death of emperor Constantine XI Palaiologos (reign. 1449 – 1453), the long-lasting agony gave way to the *post-mortem* lamentation of the Empire that was lost. Many contemporary people believed that the old prophecies stating that Constantinople would be destroyed under the reign of an emperor named Constantine whose mother's name would be Helena, just like the founder of the Empire, had come true, and that Divine will had been fulfilled. Of course, the fall of Constantinople did not signify the end of plague epidemics.[511]

Could the death of the last emperor, who happened to bear the same name as the founder of the Empire, the fall of the Empire, and the epidemics that struck Constantinople be interpreted as the fulfillment of a prophecy or the result of Divine will or wrath? In May 1456, plague conquered Constantinople more easily than sultan Mehmed (1432–1481) had done three years earlier, while Christians and Muslims

---

509 Efthychiades 1983: 214–215, 517. Further remedies in Valiakos 2019: 437, 453, 485, 1092, 1100)
510 Nicephorus Gregoras, *Byzantina Historia*, XVI, A.M. 6855 [= A.C. 1347], P. 501 (ed. Schopen 1829 – 1830: 2.798.1–5).
511 Varlik 2015: 131–159.

were lying dead next to each other on the streets of the city. Those residents who chose to run away soon realized that—wherever they went—they were still captives of the disease which spread rapidly to Thrace and had reached Bosnia and Belgrade by 1458.[512]

But even if a plague epidemic or the Fall of Constantinople was lawful and death was fated, should people have tried to change their fate even at the last moment? Would flight have been more likely to cause the wrath of God? The answer of Theodoros Agallianos (ca. 1400–1474), deacon of Constantinople and future Bishop of Médéa, is indicative [513]:

Many believe that randomness defines whether some people are going to die while others are not. Is it contrary to the plan of God when some people leave the afflicted places in order to escape death? As far as I am concerned, I think that whether one is infected by the disease or not is independent of the moral attitude, but rather depends on the physical idiosyncrasy. Why should it be a sin if someone tries to escape from the disease? If everything is predetermined, then it makes no sense to resort to drugs in the face of a disease or to grab a shield in case of war.

512 Varlik 2015: 131–159.
513 Congourdeau 1993: 21–41.

Chapter 7
Analysis of the First Pandemic

Aggregations of Epidemics during the *First Pandemic* in Byzantium

The grouping of the epidemics of the first pandemic has been the subject of historical study, whereas the criteria upon which it is based are the time of onset, the geographical distribution, and human activities (trade, military movements). Without rejecting the significance of these factors, the current study attempts to group the epidemic waves of bubonic plague that struck the provinces of the Byzantine Empire during the first pandemic, while evaluating the periodic emergence of the disease in light of the possible existence of enzootic areas in the Empire.

Epidemics are defined in terms of space, since the word 'epidemic' is often accompanied by the term 'outbreak', which is defined as "the occurrence of more cases of disease, injury, or other health conditions than expected in a given area or among a specific group of persons during a specific period. Usually, the cases are presumed to have a common cause or to be related to one another in some way. Sometimes distinguished from an epidemic as more localized, or the term less likely to evoke public panic" (CDC-Principles of Epidemiology in Public Health Practice, 2012). Outbreaks essentially share the same definition with epidemics; they are mainly related to events that take place within a limited geographic area. Let us not forget, however, that distinct outbreaks share another common characteristic, that is, they are closely related in terms of time, mainly weeks.

In the year 541 A.D., an unknown disease was about to emerge at the Mediterranean port of Egypt, Pelusium, wherefrom it spread to all provinces of the Eastern Roman (Byzantine) Empire. Because of the lack of previous sources containing accurate descriptions of the symptoms of the disease, this is believed to have been the first appearance of bubonic plague in Europe, which became known as the *Justinian Plague*, from the name of the Byzantine Emperor Justinian. By 542, the disease reached Constantinople, the capital of the Empire, and then it gradually spread westwards, until it reached Britain and Ireland.[514] The last incidence of plague of the first pandemic in the Byzantine world can be traced back in 748/749–750/751, in Syria and Mesopotamia, which had been conquered in the meantime by the Arabs.[515]

As we have seen in the previous chapter, plague periodically struck the provinces of the Empire between 541 and 751, and various attempts were made at times to group the epidemics of this period.[516] Such grouping for each region has indicated

---

[514] Russell 1976.
[515] Stathakopoulos 2004: 113–124,184–189.
[516] Le Goff 1969 (cited from Stathakopoulos 2004: 23–34).

https://doi.org/10.1515/9783110613636-010

a different frequency of outbreaks between urban, suburban, and rural areas of the Empire.[517]

In a 1969 study by the French historians Jean-Noël Biraben and Jacques Le Goff (1924–2014), the course of plague in Europe has been grouped into fifteen waves.[518] A more modern classification proposed in 2004 by the Greek historian Dionysios Stathakopoulos based on the sounder dating of historical sources, entails eighteen waves in the Byzantine Empire with an average rate of emergence every 11.6 years. The grouping and dating of the sources so far indicate that plague appeared in major urban centers of the Empire every 10–15 years. A typical example is the case of Constantinople, where six out of a total of ten appearances of the disease between the sixth and eighth century occurred at intervals of 11–17 years (every 14.2 years on average).[519] These ten outbreaks in Constantinople support the view that capital cities have always been vulnerable to the disease because of their intense commercial activity.

The table below comprises all epidemics of the period 541–750/51 in the imperial provinces of Egypt, Palestine, Syria, Mesopotamia, Asia Minor, Italy and, of course, the capital of the Eastern Roman Empire, Constantinople. The conquest of the eastern provinces by the Arabs (629–642) provides a new pathway for study, since we have Arabic historical sources to help us. They offer a description of the epidemics and allow us to compare and verify the Greek and Syriac sources (Table 1).[520]

By studying the sources, we are able to classify the outbreaks of plague by century and province. In principle, the overall recording of epidemics presents an interesting geographical orientation of the disease, since the majority of epidemics occurred in the provinces of Syria and Mesopotamia, with Italy and Constantinople along with their surroundings following suit. Based on recorded epidemics, it seems that the disease followed an upward trend in the provinces of Syria, Mesopotamia and Egypt, and gradually declined in Italy and Constantinople (Table 2).

Contemporary chroniclers often mention cities that were struck by epidemics, although such references are limited to vague geographical localizations in many instances, such 'in the East', or they simply mention the name of a province. The exact city that was affected by an epidemic is known in 53 out of a total 79 cases (Table 3).

If we follow the chronological evolution of the pandemic, we will identify different periods characterised by intense activity of the disease. Within these periods, intervals with typical sporadic outbreaks can be distinguished, which had a short duration and limited spread.

At this point it is necessary to clarify some aspects of the attempts made to group past epidemics, not only of plague, but also of other infectious diseases. These at-

---

517 Biraben 1989; Duncan-Jones 1996.
518 Le Goff 1969.
519 As mentioned in Stathakopoulos 2004: 23–34.
520 Conrad 1981; Morony 2007.

Table 1. Records of epidemics in the Byzantine Empire (6th–8th c.).

Province	6th century	7th century	8th century (until 750)	Total
Egypt	541	618, 672, 689	714, 724, 732, 743	8
Palestine	541, 542	626, 634, 639, 672	732	7
Mesopotamia	545, 557, 558	646, 669, 672,699	706, 718, 725, 732, 744, 748	13
Syria	542, 560, 567, 580, 591, 599	634, 639, 646, 687, 699	704, 706, 713, 718, 725, 732, 744, 748	19
Asia Minor	542, 558, 560, 565, 598, 599		735	7
Balkans	543, 597		711, 745	4
Italy	542, 543, 565, 571, 590	600, 608, 664, 676, 680	745	11
Constantinople (Thrace)	542, 558, 573, 585, 598, 599	618, 698	718, 745	10
Total	31	23	25	79

Table 2. Number of epidemics (6th–8th c.).

Provinces	6th century	7th century	8th century (until 750)
Egypt	1	3	4
Palestine	2	4	1
Mesopotamia	3	4	6
Syria	6	5	8
Asia Minor	6	-	1
Balkans	2	-	2
Italy	5	5	1
Constantinople (Thrace)	6	2	2
Total	31	23	25

tempts rest on a series of *de facto* assumptions; for example, when a source refers to an epidemic, the event is indeed considered to be an epidemic. The real problem in any grouping attempt lies in this very assumption: how do we define an epidemic? Nowadays, national health authorities invest much energy, time, and money in the epidemiological surveillance of various diseases. Given the existence of endemic dis-

Table 3. Plague epidemics per city (6th–8th c.).

Provinces/Cities	6th century	7th century	8th century	Total
Egypt				
Alexandria	1	1		2
Pelusium	1			1
Balkans				
Thessaloniki	1			1
Italy				
Verona		1		1
Grado	1			1
Istria	1			1
Narva	1			1
Pavia		1		1
Ravenna	2			2
Rome	3	4	1	7
Thrace				
Drizipera	1			1
Asia Minor				
Ankhara	1			1
Amida	2			2
Anazarbus	1			1
Germanicea		1		1
Emesa	1			1
Myra	1			1
Mesopotamia				
Kufa		2	1	3
Constantinople	6	2	2	10
Palestine				
Askhelon	1			1
Gaza	1			1
Jerusalem	1			1
Syria				
Antioch	4			4

Apamia	1		1
Epiphanea	1		1
Zora	1		1
Bosra		1	3	4

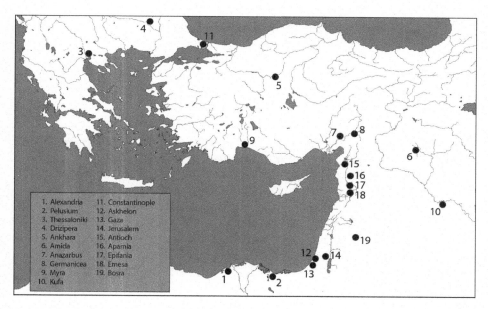

Map 1. Cities affected during the first pandemic.

eases, epidemiological surveillance constitutes the basic tool for prevention and the formulation of national health policies. When the citizens in a country are informed of an epidemic, this means that an endemic disease has been upgraded to a higher risk status, as long as no imported disease is involved. In fact, the number of cases will also determine whether a specific event is an epidemic. The abrupt introduction of a pathogen into a population, followed by the sudden emergence of multiple cases, is considered an epidemic that is classified according to its contributing factors (common source, propagated), just as it happened with COVID-19. When the frequency of cases related to an endemic disease increases, however, the classification of a nosological situation as an epidemic is a sensitive issue. If the health authorities possess an accurate dataset of the chronological sequence of events, they can classify a situation as an epidemic. A simple way to distinguish an epidemic is by counting the number of cases, which has to be higher than the average number of cases over the past five years. Public health-related issues are often approached in a distorted manner by the mass media, as a result of ignorance or sciolism, but the public opinion is nevertheless influenced. Citizens can be concerned or even scared by the news of numerous cases, but, from an epidemiological viewpoint, this news could be insignificant in terms of pure numbers—simply a normal event in the context of the periodicity of an endemic disease. Misunderstandings are common even in the modern era, the more so many centuries ago when people were ill-prepared to conceive the actual extent of an epidemic. Of course, the majority of epidemics mentioned in the sources seem to be high-risk emergency situations in line with modern standards. Unfortunately, the lack of demographic data does not allow us to approach and study each event in a comprehensive manner.

At this point, we have to ask ourselves how frequent the phenomenon of random and isolated cases of death could have been, which were never characterized as plague-related deaths by the physicians of the time and did not evolve into generalized epidemics. Could contemporary physicians have failed to make a diagnosis, at a time when plague epidemics were a common phenomenon and the clinical picture of the disease was widely distinguishable? The answer to this question is affirmative. Despite the experience gathered from frequent epidemics, a plague-related death could have escaped the attention of physicians. A telling example is Candia (Herakleion, Crete) during the Venetian rule in Crete. Its physicians had been accustomed to plague after the numerous epidemics that affected the island. Despite their accumulated experience, they failed to save the city in 1592, when two deaths that occurred within 42 days were not properly identified as plague-related. As a result, one of the most brutal urban epidemics ensued, lasting three years and causing loss of 51% of the population of the capital of the kingdom.[521]

Another issue that ought to be clarified are the so called 'epidemic waves'. The term 'wave' is frequently used not only in historical and medico-historical studies, but also in epidemiology. At present, it is considered an elementary way of expression in written and oral speech for communication purposes. According to the prevailing rationale, epidemics that emerge in different regions during a given period of time (which may be geographically neighboring and justify the spread of a disease among them) are grouped into a 'wave'. As mentioned above, historians have identified the sequence of 'waves' that affected Europe. The following example will help to understand this point.

The first period when the Empire was struck extends from 541 to 546 A.D., spanning the provinces of Egypt-Palestine (541), Syria-Constantinople (542), Italy-Asia Minor (542–543), and Mesopotamia (545–546). Based on this example, the 'first wave' lasted five years, during which nine epidemics occurred in seven different regions. For reasons of convenience and mutual understanding, the term 'wave' was employed as easy-to-use in the context of grouping. We need to clarify, however, that its use in modern studies is totally conventional. The term 'wave' was first used in the late nineteenth century, after the deadly influenza pandemic that struck Asia in 1889, in order to define the largely seasonal post-pandemic influenza mortal-

---

521 Tsiamis 2014a.

ity peaks between 1890–1894. In fact, the studies of that time associated 'waves' with the seasons of the year, distinguishing autumn, winter, and spring waves of influenza.[522]

Moreover, in the field of epidemiology, a 'wave' does not have the exact same meaning that many believe. The concept of 'wave' is related to the way a specific epidemic affects a population during one or more periods of time, that is, it is associated with one particular epidemic. The concept is essentially intertwined with the rate at which an epidemic causes clusters of incidents in a population. Its simplest and most elementary form concerns a single wave and its distinct stages: beginning, peak, and end. It is possible, however, for a 'second round' of new incidents to emerge in the same area and population upon the peak and after the beginning of its downward trend. Multi-wave epidemics are one of the major challenges for epidemiology, and entail all infectious diseases like measles, cholera, or influenza, for example.[523]

The need to redefine our terminology in order to make it compatible with epidemiological notions has to be taken for granted. An alternative term to 'waves' is the term 'cluster', which could nevertheless prove equally problematic. In epidemiology, the term 'cluster' represents several new cases seen in a particular area during a relatively brief period of time. When an epidemic spins out of control, 'cluster-like' situations involve small numbers of isolated patients, whose infectious agent does not necessarily spread to a larger number of individuals. In other words, this term also refers to the number of cases rather than to the number of epidemics.

This explanation illustrates the impossibility of invoking 'waves of epidemics' or 'waves of outbreaks'. Another term shall be proposed here, which might be employed from now on in the studies dealing with the grouping of epidemics in the course of centuries. It may be preferable to use the expression 'aggregation of epidemics' (or outbreaks), which not only avoids altering the meaning of grouping efforts, but also enhances the spatio-temporal differences in the emergence of the disease. Additionally, it causes no confusion to the reader, besides being epidemiologically unambiguous.

In the framework of epidemiology and taking historical data into consideration, 15 aggregations of epidemics and 16 isolated outbreaks have been identified for the 209-year period of the first pandemic that affected Byzantium.

Returning to the Byzantine epidemics, it becomes evident that existing sources do not allow us to identify the "individual waves" of each epidemic. Thus, we shall settle with groupings as per the notion of 'aggregations of epidemics' mentioned above. It has to be clarified that aggregations simply refer to the areas where the epidemics occurred. The lack of sources for other regions during the same period may indicate that the disease did not spread elsewhere, although we

---

522 Morens 2009; Taubenberger 2009.
523 Hoen 2015; Camacho 2018; Wunderrlich 2018.

Map 2. Byzantine provinces and Western kingdoms affected during the period 541–546.

cannot eliminate the possibility that other cases of epidemics were simply not record-ed, either due to the writer's ignorance or to the loss of sources over the centuries.

The first aggregation of epidemics occurred between 541–546 in the provinces of Egypt-Palestine (541), Syria-Constantinople (542), Italy-Asia Minor (542–543), and Mesopotamia (545–546) (Map 2).

Almost ten years after the last epidemic in Mesopotamia, a second aggregation was bound to emerge in the same province in 557, which followed nearly the same course as the first wave in reverse order, that is, Mesopotamia (557), Syria (558), Con-stantinople (558), and Asia Minor (560) (Map 3).

In the period 561–596, ten isolated outbreaks occurred in Constantinople (573, 585), Italy (565, 571, 590), Asia Minor (565) and Syria (567, 571, 580, 591). Thirty-eight years after the second aggregation subsided, during the invasion of the Avars in 598, an epidemic broke out in Constantinople that was bound to mark the beginning of the third aggregation. That aggregation spread to Asia Minor (598) and in the following year to Syria (599). At this point, it has to be pointed out that, one year prior to the epidemic of Constantinople, the disease appeared in Thessaloniki (597) (Map 4). The temporal proximity of the two epidemics indicates a possible relation between them. The fact that the epidemic of Constantinople broke out during its siege may have been a random event, and the pathogen could have invaded the city via Thessaloniki. The lack of historical sources does not allow us to be certain, but the timing can definitely not be overlooked.

Moreover, it is assumed that the epidemic that reached the Middle East in 599 is probably related to its westward spread towards North Africa (600). Therefrom, the disease reached Italy, affecting Rome, Ravenna, and perhaps Verona in the same

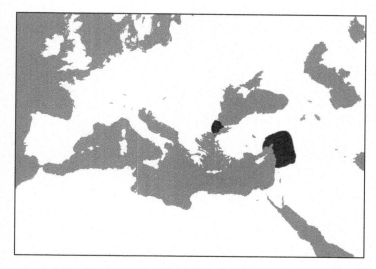

Map 3. Provinces affected during the period 557 – 560.

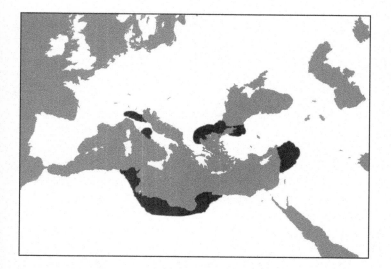

Map 4. Provinces affected during the period 597 – 599.

year. Another outbreak was recorded in 608 in North Africa, although we are unable to identify its epicenter.

In 618, the fourth aggregation emerged, starting from Egypt and moving on to Constantinople. The period when the epidemic broke out coincides with the siege of Alexandria by the Persians, at a time when the Byzantine Empire lost the province of Egypt. The disease may have been transferred to the capital via ships carrying the annual grain harvest from Egypt, just before the Persian invasion and blockade of

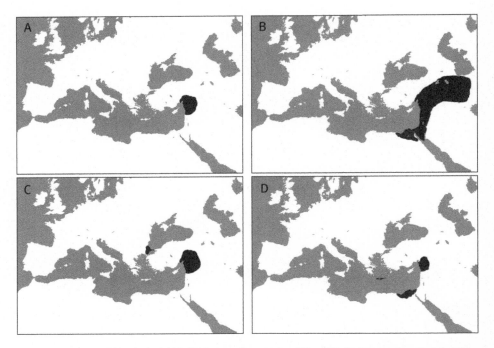

Maps 5. A. Aggregation during 646; B. Aggregation during 672–673; C. Aggregation during 698–699; D. Aggregation during 713–714.

Alexandria. An outbreak was recorded in Palestine in 626, which was obviously restricted to this specific area based on the limited historical sources available.

A fifth aggregation occurred in Palestine and Syria in 634, when two provinces were affected. The same phenomenon was recorded five years later in the same provinces (sixth aggregation in 639). The seventh aggregation of epidemics took place in Syria and Mesopotamia in 646 (Map 5 A), whereas the following aggregation was recorded in Mesopotamia in 669.

The eighth aggregation of epidemics was once again recorded in Egypt in 672, while, in the same year, it spread to Palestine, Mesopotamia, and Persia (673) (Map 5B). The following decades were characterized by sporadic outbreaks in various regions of the Byzantine world, such as Rome (664, 676), Pavia (680), Syria (687), and Asia Minor (689). The period between 698–699 marked the ninth aggregation (Map 5C), entailing the epidemics that broke out in Constantinople (698), Syria (699), and Mesopotamia (699).

Isolated outbreaks occurred between 704 and 706, and once again struck Syria (704) and Mesopotamia (706). The epidemic of Crete is a problematic case, since it might have broken out at any time between 711 and 740. The fact, however, that plague struck Syria in 713 and Egypt in the following year could justify the inclusion of the epidemic of Crete into the tenth aggregation (Map 5D), given the close trade relationship of the island with the Middle East. If this was the case, indeed, the epidemic of Crete might have taken place during the early 710s.

The eleventh aggregation coincides with the siege of Constantinople by the Arab army in 718 (Map 6 A). Plague was a trouble for both the besiegers and besieged, and the withdrawal of the Arabs was accompanied by the emergence of plague in Syria and Mesopotamia. Perhaps, however, the disease reached Syria by sea first and then moved on to Mesopotamia. In Arabic sources, that particular epidemic became known as the *Plague of Adī ibn Artāh*, named after the governor in office in Bosra ca. 713–715 A.D.

The twelfth aggregation occurred in Egypt in 724. It was obviously connected with the epidemic of 725 in Syria and Mesopotamia. The thirteenth aggregation emerged once again in the Middle East in 732, starting either from Syria or Egypt, and spreading to Palestine and Mesopotamia.

Another interesting case is the outbreak of 735 in Asia Minor. The sources refer that the epidemic also appeared in the same year on the borders of Asia Minor with Syria, during the Arab invasion. The army was decimated by an epidemic, whereas those who survived were killed by the Byzantines. The term used in Arabic sources to describe epidemics is *wabā*, which means *death* or *pestilence*. Another word encountered in Arab texts is *ta'ūn*, which is indirectly linked to plague since it describes the swelling of a part of the body (groin area).

As with the Byzantine *thanatikon*, every *ta'ūn* in Arabic sources is also a *wabā*, but not every *wabā* is necessarily a *ta'ūn*. Although the Arabic sources refer to the particular epidemic with the term *wabā*, the time of invasion in Asia Minor and the simultaneous existence of plague in Syria suggest that this epidemic belongs to the epidemic wave of 732–735 (Map 6B). It is nevertheless difficult to group the case of 735 into the aggregation of epidemics of 732.

The fourteenth aggregation occurred between 743–745 (Map 6C). The epidemic started in Egypt (743) and gradually spread to Syria and Mesopotamia (744). The same group includes the epidemic that struck Constantinople in 745. On the other hand, the outbreak of Rome (745) is problematic, since it could theoretically be related to the same aggregation, although the lack of additional data forces us to characterize it as an isolated outbreak. The pandemic of the Byzantine world ends with the epidemic of 748–750 in Syria and Mesopotamia (fifteenth aggregation) (Map 6D).

The significance of maritime and land trade routes in spreading the disease between neighbouring or distant provinces should be taken for granted. Moreover, we should not be surprised by the rapid spread of the disease across Europe. According to Maddicott's estimation, the journey of plague to England in 664 is indicative, as it traveled along the 385 miles (over 619 km) that separate Dover (Shire County: Kent, UK) from Lastingham (Shire County: North Yorkshire, UK) in a mere 91 days (4.1 km/day).

Nevertheless, we should not rule out the possibility that enzootic foci developed in some provinces of the Empire, which occasionally caused epidemics under the influence of climatic and environmental conditions.

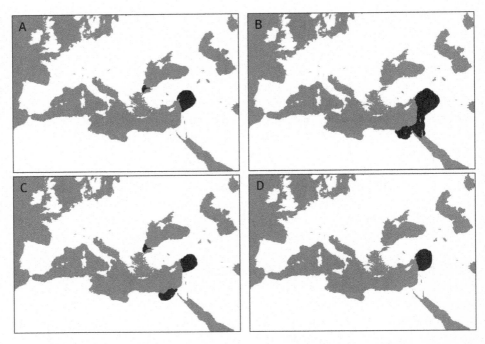

Maps 6. A. Aggregation during 718; B. Aggregation during 732–735(?); C. Aggregation during 743–745; D. Aggregation during 740–750.

Excluding those cases where the epidemic spread from province to province (Syria-Mesopotamia: 646, 725, 732, 744; and Syria-Palestine: 639, 732), there are also cases where epidemic outbreaks occurred in areas where the disease had long been absent. Although we are unable to geographically demarcate these areas, their reactivation after decades is a rather interesting phenomenon.

The epidemiological study seems to verify the theory of the permanent presence of plague in the province of Syria, since the disease appears to have affected the northern part of Syria, in the area of the 'Dead Cities', during the period 542–610 A.D. By reading the collection of poems of the Arabic writer Hassān ibn Thābit (554–674 A.D.), whose verses often evoke towns and villages struck by plague, it seems that the area bounded between Lake Tiberias, the Golan Heights, and Damascus, experienced the most epidemic outbreaks.[524] The toponyms that appear in *Dīwān*, a poem by Hassān ibn Thābit, suggest that, besides the fact that Damascus was a constant source of threat, another area north of Damascus gradually came to the fore and was apparently responsible for the emergence of various other epidemics. Verses in the poem cite these areas as susceptible to regular plague epidemics, before they came under Arab control and while they were still *ard al-Rūm*, that is,

---

524 Conrad 1994.

land of the Romans. These verses are of particular interest as they may imply that, after the aggregation of 541–546, some areas of Syria acquired permanent enzooty, which occasionally turned into epizootic and epidemic outbreaks.

Plague can be rapidly carried over long distances, either through the movement of people, or at a slower pace through epizootic rodents, which are associated with local epidemic outbreaks.[525]The case of the epidemic of 672 in Egypt—after an absence of the disease for 54 years—is strongly reminiscent of the dynamics of modern epidemics. Indicatively, we mention the case of an epidemic outbreak in the region of Oran, Algeria, in 2003, a country where no instances of plague had been reported to the World Health Organisation for nearly half a century.[526] Such phenomena can be explained on the basis of the abrupt invasion of infected animals, but also of the sudden reactivation of natural foci that had been inactive for decades. Other than human activities, the study focuses on external factors as well, such as the transfer of infected rodents via ships, which is partially true.[527]

It is often the case, however, that the role of enzootic rodents is bypassed. The enzootic behaviour of rodents allows the disease to be preserved for years. In a population of 60,000 rodents, which is a typical number for a medieval town or city, bubonic plague persists for approximately 100 years.[528] The longer this relationship among species of rodents, both resistant to the disease and non-resistant, the longer plague remains active. Naturally, this relationship has many parameters, involving the hosts and their ectoparasites, the local ecosystem (flora and fauna), and climate and topography, as well as available food and water reserves.[529]

Another interesting aspect—which probably supports the theory of enzootic areas—involves the emergence of epidemics after natural disasters. The risk of epidemics after natural disasters is taken for granted, whereas it is mainly associated with water.[530]

However, this particular risk is rather misunderstood, as it stems from the common belief—despite the lack of evidence—that dead bodies and epidemics are somehow related. Looking at modern plague epidemics, the case of the Indian state Maharashtra is an intriguing one: bubonic plague emerged in 1994 after 30 years of complete absence.[531] From an epidemiological perspective, Maharashtra is an example of special interest, given that a long forgotten focus was activated due to a change of balance between *Xenopsylla cheopis* and the little rat *Rattus rattus* following the great earthquake of 1993, which struck the same area. Within the first eight to

525 Sallares 2007.
526 Bitam 2006; Bertherat 2003.
527 Appleby 1980; Slack 1980.
528 Keeling 2000.
529 Baker 1984.
530 Watson 2007.
531 World Health Organization 1996.

ten months following a powerful earthquake, the population of rodents gradually rises to a point where this balance is disturbed.[532]

Similar cases of epidemics that broke out almost one year after a major earthquake can be identified in the first pandemic too. The well known seismic areas of eastern Mediterranean—and especially Constantinople—were struck by earthquakes on numerous occasions during the sixth century.[533]

Between the sixth and eighth century, five outbreaks appear to be the most interesting, namely those in Constantinople (558), Syria (568 and 713), Antioch (580), and Palestine (634).[534] It is interesting that four of the five records concern the provinces of Syria and Palestine, around the region of the Dead Sea, a place with a rich and well-documented history of large destructive earthquakes since 1355 B.C.[535] The northern section of the Dead Sea Rift (Lebanon, Syria) is among the main seismogenic zones. The identification of various seismic data on the elevated plateau of the region, in combination with the transformation of the area, indicate periods with violent tectonic changes during historical times.[536]

The first record of the 558 epidemic of Constantinople dates from a few months after the powerful earthquake of December 14, 557. The earthquake and the plague are well-documented in the chronicles of the Byzantine historians Agathias, Ioannes Malalas, and Theophanes.[537]

On April 2, 557 a violent earthquake struck the capital with a continuous metaseismic action for six months. The same year, a new earthquake struck again on October 6, and a sea-wave entered two miles into the area of the capital. But the absolute disaster came from a 7.0-magnitude earthquake a few months later, on December 14. The sources inform us about a third tragedy a couple of months later: the sudden onset of 'the death of the buboes'. According to Byzantine historians, countless citizens lost their lives. Once again, just like in the first plague outbreak of 542, unburied corpses filled the streets of the capital.

The second record does not provide enough information to allow us to identify the exact town, as it speaks generally of the province of Syria. According to Agapius, a major plague epidemic devastated the province in 568, linked to the disastrous earthquake in Syria (567).[538]

The previous plague outbreak occurred eight years earlier, in 560. Agapius also informs us on a third case, the plague of 580 – 581, which followed the earthquake in

---

532 McCormick 2003.

533 Glanville 1955.

534 Tsiamis 2013.

535 Sbeinati 2010.

536 Shulman 2004.

537 Agathias Myrinaeus, *Historiae*, V.3, A.C. 557, P. 145 (ed. Niebuhr 1828: 281.19 – 22); Joannes Malalas, *Chronographia*, LXVIII, O. 231 (ed. Dindorf 1831:486.23); Theophanes, *Chronographia*, A.M. 6050 [= A.C. 558], P. 195 – 196 (ed. Classen 1839: 357.14 – 18).

538 Agapius, *Historia universalis*, 175 (ed. Vasiliev 1912: 435).

the Syrian province of Antioch. Plague reemerged in Antioch 38 years later. The fourth record corresponds to the plague of 634–635. Michael the Syrian mentions a strong earthquake in Palestine, followed by 30 days of metaseismic action, which destroyed many monuments in Jerusalem.[539] Michael and Agapius mentioned that a severe plague epidemic appeared in Palestine after the disaster and spread quickly to Syria resulting in many victims. The previous outbreak appeared eight years earlier, in 626. The fifth outbreak dates back to 713 A.D. Michael the Syrian refers to a severe plague from December 712 to February 713.[540] The sources also report a devastating earthquake that took place some months prior to the epidemic.[541] In this case, the previous outbreak recorded in the province was seven years earlier, that is, in 706.

By studying the historical sources, the course and progress of the pandemic can be identified to some extent. The role of trade routes was crucial for the spread of the disease to distant provinces. However, the examination of the geographical distribution of epidemics reveals the continued presence of the disease in areas of the Middle East, mainly in the province of Syria. At first sight, plague seems to be the 'exportable product' of Syria to Mesopotamia and Palestine. Nevertheless, the disease was often 'recycled' among these three provinces, while it also affected Egypt. It is only with the help of scholarly studies and their precise dating of sources that we can define the temporal evolution and geographical distribution of an epidemic. Interestingly, Evagrius comments on the epidemics of Antioch, concluding that the disease occurred in cycles of fifteen years. It seems that foci of permanent risk existed along the borders of these three provinces. The combination of historico-geographical data captures an area that corresponds to an imaginary triangle on the top of which is Damascus, extending southeast to the city of Bosra, and southwest to the city of Fīq (the Byzantine Ippos), on the eastern shores of Lake Tiberias of the Golan Heights (Map 7).

We can conclude that the numerous outbreaks of the disease in the Byzantine and Arab Middle East could be explained in terms of enzootic presence in the area. Moreover, the existence of a natural focus on the Arabo-Byzantine border of the time should be considered very likely. As for Palestine, it appears to have suffered due to its geographic position at the crossroads of contemporary trade routes.

The monumental *Plague* manual of the WHO, which was edited in 1954 by Robert Pollitzer (1885–1968) and lists the natural foci of the disease in the 20th century, contains special references to the Middle East. More specifically:

539 Michael Syrianus, *Chronica*, Liber XI, Cap. IV (ed. Chabot 1899–1910: 4.411–413; Fr. Transl. Chabot 1899–1910: 2.413–417, especially 414).
540 Michael Syrianus, *Chronica*, Liber XI, Cap. XVII (ed. Chabot 1899–1910: 4.449–452; Fr. transl. Chabot 1899–1910: 2.477–483, especially 482); *Chronica Minora*, II, *Chronicon ad annum Domini 846 pertinens* (ed. Brooks 1904: 176).
541 *Chronica Minora*, II, *Chronicon ad annum Domini 846 pertinens* (ed. Brooks 1904: 176).

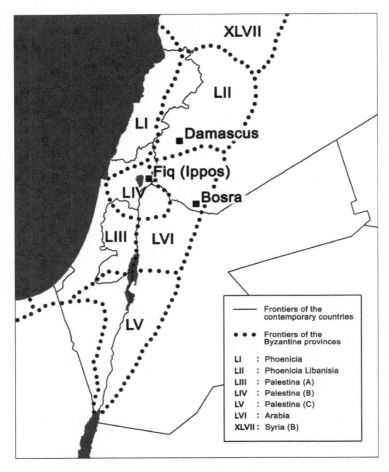

Map 7. The potential natural focus of Praefectura Anatolia, 6th c.

In Turkey, a plague epidemic in three villages on the Syrian border took place in March-April 1947. According to Erzin & Payzin, 19 persons were affected of whom 14 had axillary buboes and 5 septicaemic features. A report on this outbreak added that the infection, apparently present among the rats as well as among man, had been derived from Syria where plague used to occur in sporadic form among the desert nomads.[542]

In relation to Palestine:

... While plague cases or even limited epidemics had been met with in Palestine after the British occupation at the end of the First World War, the disease appears to have been absent during the period 1925 – 40. However, the infection, presumably imported from the Suez Canal Zone, reap-

---

542 Pollitzer 1954: 28 – 30.

peared in Haifa in 1941. Jaffa became involved in the winter of 1942–43. Plague persisted in both places and occasional instances of the disease were observed also in Tel Aviv.[543]

In Israel, the cosmopolitan *Xenopsylla cheopis* is considered as a common flea of commensal rodents. Also, the fluctuation of rodent density and flea index in the Suez Canal Zone and Sinai is of interest. We mention that the risk of plague is dependent on the increase of monthly flea index (>1) and rodent density. According to the epidemiological data in Egypt, it seems that rodents with their ectoparasites, in Port Said and the Sinai, are among the most important reservoirs for zoonotic diseases. Until 1945, plague was reported in the ports, and then spread in the mainland (high prevalence in Upper Egypt with a range of flea index 14–33 in the Aswan province). After the introduction of the insectiside Dichloro-diphenyl-trichloroethane (DDT) spraying (1946), the flea indices decreased, and the country was free of the disease. Even the low indices of the Suez Canal are of great epidemiologic interest. According to available data until early 2000, in Port Said and the Custom area, the higher flea indices occurred in September (flea index >2). [544]

Pelusium, 541 A.D.: Epidemiological or Historical "Destiny"'?

The recounting of the terrible pandemic highlights particular and significant epidemiological questions with historical connotations. According to Procopius, Pelusium reportedly holds the unfortunate preeminence as the city where the epidemic broke out for the first time. In 542, the disease swept through the capital of the Empire, which apparently received alarming messages—at regular intervals—from the provinces in the South and East. The confidence of Procopius regarding the 'responsibility' of Pelusium leaves no margins for dispute, as he states that "it started from the Egyptians who were moving towards Pelusium".[545] In turn, an attempt to interpret such confidence leads to the obvious question: How long had the epidemic been raging in Pelusium?

Based on the study of the sources, if the epidemic reached Alexandria two months later, then everybody was well aware of where the epidemic had started from. Thus, it is now apparent why—even in remote Constantinople—Procopius received information one year later, suggesting that Pelusium was the 'source of evil'. Obviously, the assurance of Procopius that the disease started from Pelusium indicates that the problem of the particular city was well known in the surrounding areas and, perhaps, everybody thought that the disease would be limited thereto. Based on the general pattern of spread of an epidemic, the epidemic in Pelusium

---

543 Pollitzer 1954: 28–30.
544 Pollitzer 1954: 28–30; Hussein 1955: 29–48; Riffaat 1981; Shoukry 1986; El-Kady 1998; Krasnov 1999; Allam 2002.
545 Procopius, *De Bellis, De Bello Persico*, II.22.6 (ed. Haury 1905: 250.13–14).

should have already been going through its phase of acme by the time it reached Alexandria.

Belonging to a typical infectious (progressive) disease, the outbreaks must have been chronologically distributed over a period far longer than the variation range of the incubation period of the disease. The resurgence of outbreaks in the initial phase of an infectious epidemic reflects the progressive increase of infectious sources and, consequently, the corresponding increase of active contacts, whereas the decline of outbreaks during the final phase reflects the establishment and prevalence of collective immunity. The outbreak of an epidemic is usually due to the introduction of the infectious agent, through an initial 'primary' outbreak (in rarer cases, more than one primary outbreaks occur simultaneously). Therefore, the rapid isolation of the particular outbreak (or outbreaks), as well as of all individuals who have been in contact with the individual(s) (that is, carriers in the stage of incubation or asymptomatic carriers), can possibly contribute to stemming the rate of the epidemic.[546]

When examining infectious epidemics, an epidemiological parameter is used, which characterises infectious diseases and goes by the title of 'serial interval' or 'generation time'. The serial interval begins just like the incubation period at the time of infection of the individual, but, whereas the incubation period ends when the first symptoms appear, the serial interval ends when the infectivity of the new outbreak is at its peak (which is usually apparent before any clinical manifestations). This parameter is significant, since it does not only involve clinical, but also subclinical cases that contribute to the spread of the infectious agent and, consequently, of the epidemic.[547]

The view of Procopius that the disease started from the sea and spread towards the inland seems to correspond to reality. We could use the information provided by sources in order to redraw the course of the disease from Egypt to Constantinople. We now know that it all started in Pelusium, which was misfortunate enough as to go down in history as the source of disaster. However, one question remains unanswered: Why did it all start in Pelusium? Consequently, many more spontaneous questions arise. Why did plague 'choose' Pelusium and not Alexandria, or another city right from the outset? Was it 'misfortune' or 'coincidence'? Could it be the case that Pelusium was the ideal city to host plague at that particular time? Pelusium was a simple link in a chain that started farther away. Was another city to be 'held accountable' for what happened in Pelusium? All these questions are summarised in one important question: Ultimately, what was the real gateway of plague to Europe during the sixth century, and where did the disease come from?

Based on the reference of Evagrius that "rumor had it that it (the disease) came from Ethiopia" in his *Ecclesiastical History*, as well as on the existence of natural foci in the interior of Africa until today, various researchers have hypohtesized that the

---

546 Halloran 2001.
547 Halloran 2001. See also Centers for Disease Control and Prevention 2012: 36–42.

disease started in Central Africa.[548] The areas through which it purportedly reached medieval Ethiopia correspond to modern Uganda, Kenya, and Zaire. To this day, Central Africa is host to natural foci of the disease, many of which are under ongoing monitoring by the services of the World Health Organisation. However, the first officially recognised bubonic plague epidemics in Central Africa were recorded in the late nineteenth and early twentieth century. More specifically, the disease first appeared in Uganda in 1896. Of particular interest is the fact that the epidemic of 1896 broke out during the drilling of a rail network in the country. During the same period, the company responsible for the project was also carrying out works in India, wherefrom it transferred workers and equipment to Uganda. In Kenya, plague appeared in 1902, whereas in late 1920s it made its first appearance in the areas of the Belgian diamond mines of the Democratic Republic of the Congo (former Belgian Congo).[549] These three countries could also include the former German colony of East Africa (modern Tanzania), where the German medical missions of the late nineteenth century led by Robert Koch (1843–1910) had identified endemic foci on the border with Uganda, in the areas of Uhehe, Iringa, the Kagera River, and the Lake Victoria.[550]

Nevertheless, it is hard to determine how old an endemic focus is, since the data of the WHO are no older than the early nineteenth century. According to modern data, the most dangerous areas of Africa with proven enzootic foci are located in the provinces of Ituri, Rimba, Bas Uele, Haut Uele, and Rethy in Democratic Republic of Congo, as well as in the province of Nebbi in Uganda. However, we are unable to determine the enzootic foci before the nineteenth century.[551] Apart from the theory of African origin of the disease, India has also been indicated as a candidate. In any case, the existence of similar foci in India leaves space for the theory of an Indian origin of plague.[552]

The Port of Clysma as a Possible Gateway of the Disease

The geographical position of Pelusium lies in the area of the modern city Tell el-Farama, in the eastern delta of the Nile estuary, thirty kilometers southeast from the Mediterranean entrance to the Suez Canal, at Port Said. The port of Pelusium was overshadowed by Alexandria, wherefrom ships set sail for the long voyages of that time. However, the role of Pelusium was far from unimportant, given that the goods of Egypt—mainly cured meats, oil and textiles—were transferred there on

---

548 Braton 1981; Biraben 1975.
549 Pollitzer 1954: 38–46.
550 Pollitzer 1954: 38–46.
551 World Health Organization 2003; Bertherat 2005.
552 Krishnaswami 1972; Srivastava 1978.

their way to the cities of Palestine and Syria.[553] At the time when the epidemic broke out, the Byzantine Empire was at its peak, with the Mediterranean having become the Byzantine *mare nostrum*. The hypothesis that the disease was transferred to Pelusium from another port of the Mediterranean seems rather unsound. Should this hypothesis be true, however, then we are automatically driven to conclude that an endemic or epidemic focus was already existent within the Byzantine territory, probably located among the coastal cities of the Mediterranean. It would also make sense that outbreaks of plague along the Mediterranean coastline occurred before 541, but this assumption is not supported by any reference among currently available sources.

This event did not escape the attention of researchers, who have supported the theory that the disease spread from the Red Sea to the Mediterranean.[554] The hypothesis of overland spread of plague to Pelusium urges us to study the commercial network of the Egyptian mainland.[555] It is necessary to refer to the geography of the area, for the vulnerable position of Pelusium to be highlighted. Pelusium was not a port in the modern sense of the word, since it was located 4–5 kilometers away from the Mediterranean coastline. It was accessible only to particular types of small vessels. Perhaps these small vessels on the Mediterranean coastline received goods from the bigger ships and transported them to Pelusium through the canals. Moreover, just like its name suggests (Pelusium < Greek πηλός, meaning *clay*), the city had been built upon the marshy and muddy areas of the delta, notwithstanding the implications of such a feature for the public health of a city.

Pelusium was located in one of the eastern branches of the Nile delta, an area which was well-known since Roman times as *Ostium Pelusiacum*. The ancient delta of the Nile consisted of seven branches, the flow of which has now been changed or disapperared, as was the case in the East, due to the construction works of the Suez Canal. During the Roman times, the river delta comprised a complex network of canals which extended around the Grand Canal (modern El-Baqar), which, in turn, connected Rosetta with the West and Damietta with the East, along the Mediterranean coastline. With El-Baqar being the central axis, smaller canals spread radially and often intersected each other, thus creating a complex network that connected various stations, from Cairo to the Mediterranean. According to the narration of Procopius in his work *Buildings of Justinian*, the Nile did not reach as far as Alexandria, but rather stopped at the small town of Haireu. A new canal was created there, which connected Alexandria to Haireu, but it could only be traversed by one specific type of small vessel—the so called διαρήματα (*diarêmata*).[556]

In Roman times, Trajan's Canal was an equally significant waterway, as it linked the canals of the eastern delta of the Nile with the Red Sea. Trajan's Canal, which at certain points reached 100 feet in width and 40 feet in depth (30 and 42 meters, re-

553 Milani 1977: 216.
554 Sarris 2007: 121; McCormick 2007.
555 Tsiamis 2014b.
556 Procopius, *De aedificiis*, VI.1.1–4 (ed. Haury 1913: 171, esp. 171.16 for the term διαρήματα).

spectively), went through, and connected, the cluster of the Bitter lakes, whereas this particular idea was also used in the design and drilling of the modern canal. However, the fluctuations of the level of the Nile destroyed the canal, and the negligence of the authorities concerning its maintenance could make it non-navigable. Archaeological excavations have brought to light the remnants of two canals, upon which two new ones were later built, namely *Tell-el-Fadda* and *Tell-e-Luli*, which started at the point where the Trajan's Canal joined the Nile and were heading to Pelusium. The dating of river shells with carbon ^{12}C indicates an age of approximately 1925 ± 100 years, which coincides with the change of flow in the eastern Nile at the time of Trajan, as a result of the works to extend the canals.[557] These structures in the eastern delta were aimed at the faster transference of goods from the Red Sea to the Mediterranean, by bypassing the main volume of the river for reasons that will be explained later.

During the Byzantino-Persian wars, a peculiar war took place, where both the Byzantines and the Persians were engaged in a struggle for influence in the kingdoms of *Axum* (modern Ethiopia) and *Himyar* (modern Yemen). In 525, the rise of Jewish kings towards the land of the Himyarites (Homerites) and the ensuing massacres of Christians, prompted the Axumites—who were also Christians—to intervene militarily with the help of the Byzantines.[558] This interference led the Persians to get involved in the conflict by assisting the Judean Himyarites. At the borders of Byzantium and Persia, the customs-station of Ktesiphon remained closed, whereas in the East the Sassanid Persians controlled the trade of the Persian Gulf, part of the trade of the Indian Ocean, as well as the land route of the Silk Road. The sea, however, was not completely under their control, so the Axumite (on behalf of the Byzantines) and the Himyarite merchants were able to reach Ceylon (modern Sri Lanka, called Taprovani by the Byzantines) or southern India, as testified by the numismatic findings of *solidi* (gold coins first issued by the Romans) that date back to the period between Theodosius II (reign. 408–450 A.D.) and Heraclius.[559] In general terms, the Empire had no reason to be alarmed by the Kingdom of Axum in the South, since it enjoyed the fruits of the diplomatic undertakings of emperor Constantine I in the fourth century A.D. The Axumites who lived in the country that was located at the outermost point of the "ocean" served as an alternative commercial solution, whereas they were able to challenge the Persian Empire in the South whenever deemed necessary, as well as to defend the entrance to the Red Sea and, of course, to guarantee the security of the Byzantine commercial activity in the area.[560]

The Kingdom of Axum was a retail commercial hub of goods from Central Africa and India. Trade ships transported fragrant plants, spices, herbs, precious and semi-

---

557 Sneh 1973; Stanley 2008.
558 Laiou 2007: 36; Hendrickx 2012.
559 Laiou 2007: 36.
560 Casson 1989: 17–18.

precious stones, ivory, and silk fabrics from China. [561] Every journey from Africa or Arabia towards India involved numerous interim stations, which were also used by the Arabs in the following centuries. According to the descriptions of the 10th–century Arabic geographers Ibn Hawqal (fl. 977 A.D.) and Muqaddasi (45–991 A.D.), a trip from Yemen to India would initially take place along the coast of the Arabian Peninsula and across the Persian Gulf, with stopovers at the ports of Baçra, Shatt al-Arab, Siraf, Çuhar, and Dayboul, before the ships reached the coast of West India.[562]

The Byzantine ships that carried the goods of the Axumite and Christian Himyarite merchants entered the Red Sea, heading towards the ports of the eastern coast of Egypt, in order for the caravans to approach the Nile through the old Roman road *Via Hadriana*, which connected the major ports of Berenice and Myos Hormos with riparian Antinopolis.[563]

Another station in the Red Sea was the port of Iotabe, the exact geographical location of which remains unidentified until today, due to the contradictions in existing historical references. In essence, Myos Hormos was the last accessible port, since the river abruptly turned shallow beyond this port, thus making navigation difficult.[564]

Nevertheless, there were two ports in the Sinai Peninsula which served the commercial needs of Palestine and Syria: Ailana (modern Aqaba) and Arsinoe. Trade contributed to the prosperity of the commercial stations of the Red Sea, but not of Arsinoe, which was ultimately abandoned due to strong southeastern winds in the area. As a result, the need to create a new port emerged, about 5–7 miles (8–11 km) westwards and towards modern Suez, that is, the Port of Clysma (Al-Qulzum) which is often, but wrongly, equated with Arsinoe.[565] After dredging works in the old and unwieldy Roman port of Clysma took place, a new pole of attraction for merchants was established, which gradually overshadowed the other stations in the Red Sea and represented the ending point of the journey from the 'outermost ocean'. Clysma over time evolved into a port with great commercial activity that contributed to its prosperity. The choice of Clysma was not accidental, given its location that essentially distinguished it as the entrance to the Trajan's Canal. Clysma and Pelusium were in direct contact with the canal (*Amnis Traianus*), as well as with the smaller passages of *Tell-el-Fadda* and *Tell-e-Luli*.

Trajan's Canal was directly dependent on the level of the Nile.[566] The fluctuation in the level of the Nile must be considered. Between March and May it reaches a minimum threshold, its peak being around mid-September, and it starts dropping again in November until December. These fluctuations differ from one year to another, the

---

561 Casson 1989: 17–18.
562 Miquel 1977: 131–145.
563 Sidebotham 1997; Sidebotham 2002.
564 Young 2001: 75–78.
565 Sidebotham 1991.
566 Collin 2007: 19–46.

volume of water ranging from 36 x 10^9 m^3 to 7 x 10^9 m^3.[567] Following the flow of the 1,160 miles (6,695 km) long river from Central Africa, we find that its level is at its lowest point in May, and starts rising in the period between June and August. These fluctuations are believed to have been an important factor in the history of plague and Pelusium. If we accept the theory that the time when the epidemic broke out in Pelusium was mid-July 541, and in conjunction with the facts that the level of the river went up at the same time and Trajan's Canal was in operation, as well as that the disease appeared in Alexandria in mid-September of the same year, we could historically and geographically link the route between Pelusium-Alexandria with the network of canals and the port of Clysma as the gate of the pathogen.[568]

One could arguably reject this hypothesis, by proposing that the epidemic may have been transferred directly through the Nile, thus reducing the role of Clysma. However, it was obviously time-saving for goods headed to Palestine and Syria to be transferred from Axum to Clysma through the sea rather than the Nile. The reason why we stand in favour of this view rests on the difficulties of navigating through the Nile. As strange as this may seem, a journey through this river was much slower. For example, the distance from Sudan to the Mediterranean is 960 miles (1,545 km), while the distance between the Aswan Dam and the Mediterranean is 745 miles (1,200 km). The rise and decline of the level of the Nile, combined with the flow of the water, created new and 'uncharted' reefs in the river bed; therefore, already since the Pharaonic time, ships would only travel during daytime, making sure that they reached a station on the banks of the river before nightfall.[569] The ships could then remain at these stations for one or two days, awaiting the caravans that in turn, struggled to reach the ships. Finally, particular spots of the Nile were not navigable at all, which meant that the ship had to be towed from the shore with the help of animals. It is estimated that a vessel sailing along the Nile could cover a distance of 25 to 43 miles (40 to 70 km) per day.[570] Consequently, if we accept that each ship would cover approximately 34 miles (55 km) per day, excluding the possibility of further delays, the distance of 960 miles (1,545 km) separating Sudan from the Mediterranean would take about 28 days. The transfer of goods from Axum to Clysma and from there to Pelusium was successful in almost half of annual efforts. It makes sense that the traders of Central Africa and Axum preferred to transfer their goods to the ports of the Red Sea through the maritime routes. Moreover, in line with Procopius' report, the route between Axum and the border city of Elephantine, on the Egyptian border of the Byzantine Empire, requires thirty days of walk for a wayfarer.[571]

---

567 Abu-Zeid 1983; Fraedrich 1991; Eltahir 1996; Mikhailova 2001.
568 Tsiamis 2009.
569 Nachtergael 1988: 19 – 54.
570 Nachtergael 1988: 19 – 54.
571 Nachtergael 1988: 19 – 54.

The proposed choice of Clysma is reinforced by yet another historical event. In the sixth century, the ports of Myos Hormos and Berenice had been abandoned and the *Via Hadriana* was destroyed.[572] The archaeological and anthropological findings in the area of the eastern desert of Egypt bear witness to a gradual gathering of nomads since the third century A.D., who launched raids against the ports of the Red Sea and the Nile.[573] According to the most recent report on the operation of Berenice, which dates to 524–525, the defense of the city was vestigial for almost a century; this is also witnessed in the *Notitia Dignitatum*, where the departure of the *Legio I Valentiniana* in the fourth century is recorded.[574] At the same time the position of Clysma was upgraded—the city even became host to the λογοθέτης (*logothetēs*), a high ranking official who visited Axum once a year acting as ambassador.[575]

The massive withdrawal from many hubs on the coastline of the Red Sea and the Nile denotes the fear of nomadic raiders. We assume that this fact contributed to the qualification of the particular route from Axum to Clysma as preferable. The feasibility of navigation beyond Clysma through Trajan's Canal is often called into question.[576] In fact, between the second and the fourth century, the city and its surrounding areas were uninhabited until 383, for which year we have the testimony of Egeria (or Etheria).[577] At that time, it is only reasonable that the imperial authorities would have abandonded the wider region of the Sinai and, thus, the canal. Nevertheless, due to the growth of trade and historical conjunctures, Clysma resurfaced as a major port, and the canal is believed to have been reopened.[578] Gregory of Tours offers a remarkable clue, in an attempt to explain the Biblical Exodus. Citing the testimony of contemporary travelers, he describes the existence of canals from the Nile's Babylon (Cairo) to the Bitter Lakes of Sinai.[579] He also adds that the location of Clysma had not been chosen due to the fertility of the area, but rather because this port constituted both the end of the canal and the stopover for goods transported from India to Egypt.[580] Moreover—still according to Gregory—these structures received the necessary water from the Nile, whereas the flow went from the West (Egypt) towards the East (Sinai). This is a clue that strengthens the view that the operation of the canal was inextricably linked with the fluctuations of the Nile.

---

572 Sidebotham and Zitterkopf 1996.
573 Ward 2007.
574 Ward 2007.
575 Ward 2007.
576 Sijpesteijn 1963.
577 Egeria, *Itinerarium* Cap. 7.1–3 (ed. Maraval 1982: 152–154 for the Latin text, and 153–155 for a French translation). This part of Egeria's *Itinerarium* is reconstructed from the 11th-century *Liber de locis sanctis* of Peter the Diacon (on this point, see Maraval 1982: 43–44, 56–57n1, 104). On the use of Egeria's text by Peter the Diacon see also Mayerson 1996. On Clysma, see Maraval 1982: 105, 107–11. Also Wilkinson 1981: 179–180.
578 Tsiamis 2009.
579 Gregorius Turonensis, *Historiarum Libri*, I.10 (eds Krusch and Levison 1937: 11–12).
580 Gregorius Turonensis, *Historiarum Libri*, I.10 (eds Krusch and Levison 1937: 11–12).

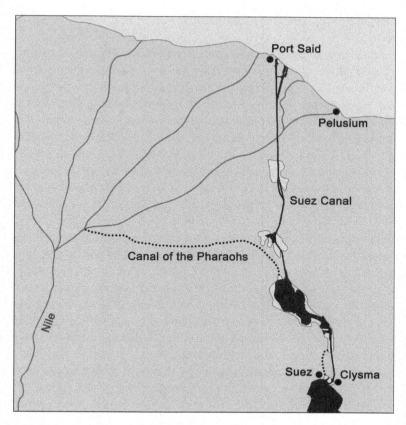

Map 8. From Clysma to Pelusium and from Suez to Port Said: the evolution of connection between Red Sea and Mediterranean Sea.

In any case, we should not neglect the fact that Clysma and Pelusium were directly interconnected through the Bitter Lakes and the Sinai oases (Map 8). The danger of raids by the Saracens of the Sinai Peninsula against the caravans was omnipresent and frequent. During the first Byzantine centuries, the term "Saracen" was used to designate the Arab tribes of the Sinai Peninsula. In the next centuries, the term was extended to any person (Arab, Turk, or other) who professed the Islamic faith. Our sources also reveal an incapability to protect the overland-routes of the peninsula from the 'scourge' of the Saracens. Apparently, despite the short duration of the overland route, traders preferred the canals for safety reasons. Archaeological findings reveal that the settlements of the peninsula were defensively over-fortified, a fact that is obviously bound together with the abandonment of the region of Sinai by the state, after the redeployment of *Legio X Fretensis* to Aelana, in order to defend the *Via Nova Traiana* that led to Syria.[581]

---

581 Parker 2002: 77–82.

The Molecular View of the First Pandemic

The study of the recursion of the birth and course of *Yersinia* with the help of the 'molecular clock hypothesis' is equally interesting. According to this theory, when the sequence of variety and divergence has a relatively constant rate, then a common ancestor of all microorganisms can be figured out.[582] *Escherichia coli* and *Salmonella enterica* are believed to have emerged from a common yet unknown ancestor. Presumably, *E. coli* was detached from the common bacterium *E. coli-S. enterica* 140 million years ago.[583] Accordingly, a new bacterium developed out of *E. coli* 18 million years ago, which is the ancestor of *Yersina enterocolitica*. It is estimated that approximately 2 million–400,000 years ago, the strain *Y. pseudotuberculosis* developed from *Y. enterocolitica*, so that we are now confident that *Y. pestis* is a clone of *Y. pseudotuberculosis*.[584]

All calculations were carried out through mathematical models on the basis of the rate of mutation of *E. coli*, which amounts to 300 generations per year, as well as of the annual polymorphism of *E. coli* and *S. enterica*, which is in the order of 6×10^{-9} and changes in later bacteria at 3.4×10^{-9}.[585] These mathematical calculations enable us to define two confidence time intervals, within which the emergence of a new bacterium out of its ancestor took place. The confidence interval when *Y. pestis* separated from *Y. pseudotuberculosis* ranges from 20,000 to 1,500 years.[586] These are obviously mere calculations rather than definite facts.

What is striking about these calculations is the minimum time limit of 1,500 years. Assuming that *Y. pestis* separated from *Y. pseudotuberculosis* some 1,500 years ago, this event is dated to the sixth century A.D., that is, the century when the First Pandemic occurred. If this is true, then we are witnessing the birth of a new organism, which was not virulent in the beginning (with the human-avirulent strain *Y. pestis 91001* being its representative sample), but soon evolved into a highly aggressive bacterium that is directly dependent on a host, thus causing the massive death of hundreds of thousands of immunologically vulnerable people.

But where could the bacterium have reached Byzantium from? This question essentially raises another major question relating to the origin of the first pandemic. The narrative of Procopius who supports the African origin of the disease, the reference to 'Ethiopia' by Evagrius, as well as the existence of the 'Angola' African strain, have long been the basis of the theory that the first pandemic originated in Africa. With the help of historical sources and archaeological remains, we were able to reconstruct some of the commercial routes of the area. Based on the above, it appears that the commercial traffic in the Red Sea was high. It also appears that the origin of

---

582 Ochman 1987.
583 Ochman 2000.
584 Paradis 2005.
585 Whittam 1996.
586 Achtman 1999.

the bacterium of plague is doubtful, with some historians proposing that *Y. pestis* originates from Central Africa, and others suggesting India. The mention of India in historical sources is not necessarily indicative of the Asian country, given that it is impossible to define with precision what each writer meant by 'India'. After all, the geographical conception of the world in medieval times was quite different from modern understanding. Procopius' report is indicative: in his *History of the Wars* he makes special reference to the Indian and the Ethiopian traders, whereas in *On Buildings* he replicates one of the prevailing views of his time, according to which "the Nile is the river starting from the land of the Indians and leading to Egypt".[587]

Before looking into the origin of the pandemic, let us briefly remember the main classifications of the bacterium. Firstly, we need to clarify that *Y. pestis* has been divided into biotypes, based on the ability of the bacterium to ferment glycerol and arabinose, as well as on the reduction of nitrate to nitrite. Based on these properties, the following biovars can be distinguished today (Table 4):

Table 4. Classification of *Yersinia pestis* biovars based on biochemical properties.

	Antiqua	Mediaevalis	Orientalis	Pestoides/Microtus
Fermentation of Glycerol	+	+	-	+
Reduction of Nitrate	+	-	+	+
Fermentation of Arabinose	+	+	+	-

Moreover, according to genetic diversity studies, an additional biovar has been proposed, that is, the Intermedium biovar.[588] Molecular studies allow us to further classify the strains.[589] A usual classification is based on the analysis of the Single Nucleotide Polymorphism (SNPs), grouping modern isolates into five branches (Figure 2)[590].

- Branch 0: Pestoides
- Branch 1: Orientalis (1.ORI) and Antiqua (1.ANT) isolates from Africa
- Branch 2: Medievalis (2.MED) and Antiqua (2.ANT) isolates from East Asia
- Branch 3 & Branch 4: modern representatives of *Antiqua*

Branches entail sub-branches and clusters which are geographically defined. Four populations have been grouped into Pestoides/Microtus (0.PE1–0.PE4).[591] The non-human pathogens of biovar Microtus spread all over Central Asia, China, and

---

587 Procopius, *De aedificiis*, VI.1.6 (ed. Haury 1913: 172.5–6).
588 Li 2009.
589 Radnedge 2002; Dai 2005.
590 Chain 2006; Drancourt 2012.
591 Morelli 2010.

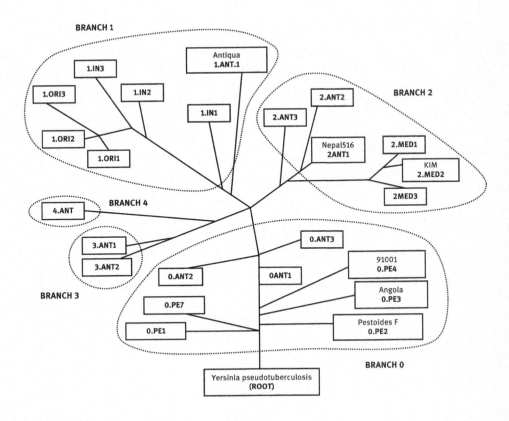

Figure 2. Simplistic phylogenetic tree of *Yersinia pestis* by SNPs analysis (the distances between the strains are symbolic).
Based on the following studies (in chronological order): Achtman 1999; Parkhill 2001; Deng 2002; Achtman 2004; Drancourt 2004; Song 2004; Chain 2006; Haensch 2010; Morelli 2010; Bos 2011; Harbeck 2013; Wagner 2014; Duchemin 2015; Drancourt and Raoult 2016; Andam 2016; Kutyrev 2018.

Mongolia.[592] Of course, molecular studies progress and new genomes are analyzed and constantly enlisted. Apparently, the global geographical distribution of the bacterium leads to the constant inclusion of new data into classifications. For example, the data gathered in the context of a major study of 178 strains from 23 plague foci in Russia resulted in the introduction of ulegeica (0.PE5), tibetica (0.PE7), and qinghaica (0.PE10) into the pestoides, whereas it concluded that the branch 2.MED is divided into the Caucasian–Caspian and Central Asian–Chinese branches.[593] Notably, the

---

592 Li 2009.
593 Kutyrev 2018.

existence of *Y. pestis* strain 91001, which was isolated from the *Microtus brandti* in Inner Mongolia, is human avirulent, but lethal to rodents. The ancestral *Y. pestis* might have been virulent only to rodents in the beginning, with some of its strains acquiring new genes and managing to overcome the limitations of their initial host, thus broadening the range of their hosts.[594]

More recent studies examining numerous strains reveal the constant presence of the bacterium in Asia prior to the first pandemic. The study of 563 samples from skeletal remains (bones and teeth) found in Russia, Croatia, Estonia, Germany, Hungary, Latvia, and Lithuania, which are believed to date from the late Stone Age (ending between 8,700 and 2,000 B.C.) and the Early Bronze Age (c. 3000 – 2100 B.C.), revealed the presence of a microbe containing parts of the sequence of *Y. pestis*. These findings leave open the possibility of the introduction of the microbe into Europe during the migration of populations 4,800 years BP. We still do not know how the disease was transmitted, but it has been argued that the bacterium reached the European continent sometime between 3,000 – 5,000 BP. In fact, it is estimated that after its initial introduction into Europe, the microbe could have gradually returned to Central Eurasia during the Bronze Age (3100 – 300 B.C.).[595] A similar study of 101 samples from skeletal remains found in Asia and Europe covering the period 2,800 – 5,000 BP also identified sequences of *Y. pestis*. The absence of crucial virulence genes is particularly important. It is extremely difficult to reconstruct the pathway of the bacterium towards Europe, but it appears that it was already present since the late fourth millennium B.C.[596] The first estimate, however, is that these strains have no direct relation to the one that affected the Byzantine Empire during the sixth century A.D.

The studies concerning the strain that caused the first pandemic are of particular interest. The comparative analysis between 233 modern genomes and the genomes of the two pandemics highlighted the relationship between the genomes of branches 0.ANT1, 0.ANT2 and 0.ANT5 and the strains found in the Xingjiang region in north-western China, or in Kyrgyzstan.[597] The study of 56 *Y. pestis* strains from three highmountain foci of Kyrgyzstan (Tien Shan, Alai, and Talas) produced similar results. It was determined that the strains belong phylogenetically to 0.ANT2, 0.ANT3 and 0.ANT5. In fact, the strains of branches 0.ANT2 and 0.ANT3 had also been found in China, while the 0.ANT5 phylogenetic branch was identified for the first time.[598] Furthermore, the basis of all strains of the first pandemic seems to have originated in the genome (second–third century A.D.) of the Tian Shan mountain range in Central Asia.[599] The variety and spread of the strains is also indicative of the great flex-

---

594 Song 2004.
595 Valtueña 2017.
596 Rasmussen 2015; Rascovan 2019.
597 Wagner 2014; Keller 2018.
598 Eroshenko 2017.
599 Harbeck 2013; Rasmussen 2015; Cui 2016: 185; Rascovan 2019.

ibility of the genomes, which are able to adapt to new environments. The first indications, however, show that this area could have been the place of origin of the branch 0.ANT as the ancestor of the high virulence strains. With the help of phylogenetics, the theory of the African origin of the pandemic loses ground.[600] This however does not negate the possibility that the strain was transferred from Asia to India and Africa through the historically recorded trade routes and historical events mentioned above. Central Asia appears to have been the place where the journey of the deadly strain that swept Europe began, while the branch that caused the *Plague of Justinian* seems to belong somewhere between groups 0.ANT1 and 0.ANT2. The data available so far indicate that both are far from those of the second and third pandemic.

As far as the theory of African origin is concerned, an interesting aspect can be identified. Studies point to a close relationship between the Angola strain and the Pestoides group, thus rendering it one of the most ancestral strains within the *Y. pestis* lineage. Analyses place the Angola strain on the 0.PE3 branch, fairly close to the Pestoides group isolates F (0.PE2a) and 91001 (0.PE4), so that this specific African strain is now considered as a representative of the microbe's ancestral lineage.[601]

---

600 McCormick 2007.
601 Eppinger 2010.

Chapter 8
Analysis of the *Second Pandemic*

Kaffa-Trebizond-Constantinople: Re-Evaluation of de' Mussi's Narrative

As we have already mentioned, the Byzantine historiography lags behind the historiography of Western Europe as far as the recording of information on the *Black Death* epidemics is concerned, thus hindering our efforts to portray the evolution of plague in the Byzantine territories. The account of Gabriele de' Mussi (c. 1280 – 1356) remains our main source for studying the onset of the *Second Pandemic*. Nevertheless, a thorough reading of the narrative obscures the mystery surrounding the reemergence of the disease rather than shedding light on it. It should be noted that—unlike the prevailing view—the notary Gabriele de' Mussi of Piacenza was not actually present in Kaffa during the siege of the city; instead, his narrative of the events may rely on the gleaning of the numerous accounts of eyewitnesses relating to the dramatic siege. de' Mussi presumably completed his work between 1348 – 1349, and thus retrospectively recounts the events in Kaffa, as well as the overall situation in Italy at the time of the epidemic.[602] The narrative comprises two specific events that deserve particular attention, namely the transmission of the disease from the Mongolian military camp to the interior of Kaffa through the corpses that were hurled over the city walls by trebuchet, and the spread of the epidemic to the Mediterranean by the Genoese galleys.

Regarding the means of transmission of the disease to the residents of Kaffa, the account of de' Mussi, according to which the Mongols threw the carcasses of the dead hoping that the Christians would die due to the intolerable stench, reflects the general medical perception of the time on disease transmission.

The prospect of disease transmission through the contaminated air emitted by corpses seemed a rather rational hypothesis based on contemporary standards. The existence of fleas on dead corpses seems a possible scenario, besides modern microbiology suggests that *Y. pestis* can survive on the surface of objects under specific conditions and maintain its infectivity.[603] Another hypothesis, which has been put forward in relation to the transmission of the disease, concerns the transmission of *Y. pestis* via rodents. It is speculated that infected rodents from the Mongol camp transmitted the infectious agent to the rodents of Kaffa. This speculation seems rather impossible, however, for reasons related to the overall behavior of rodents, as well as to the way in which medieval sieges were carried out. More specifically, the shooting range of trebuchets used for defending medieval cities during the four-

---

602 Tononi 1884.
603 Rose 2003.

https://doi.org/10.1515/9783110613636-011

teenth century fluctuated between 800 to 1,000 ft (250 and 300 meters). This practically meant that the camp of the besiegers had to be set up at a greater distance.[604] Therefore, the rodents that possibly existed in the Mongol camp would have abided to their primal instinct, and their behavior must have been similar to that of modern fossorial rodents, which do not draw away from their nest for more than a few dozen meters.[605] As far as the survival of *Y. pestis* in soil is concerned, the most recent studies have shown that the bacterium can persist and retain its virulence up to 40 weeks.[606]

The timeline of the siege is verified by the reports of the historians Ioannis Kantakouzenos and Nikephoros Gregoras, although they never mention the event of throwing carcasses in the city. Furthermore, it is impossible to specify the exact time when the events would have occurred. This temporal discontinuity is inextricably linked to the second vague aspect of de' Mussi's account, namely the voyage of Genoese galleys in the Mediterranean and the spread of the disease. Given that the ships arrived at Messina in October 1347, the actual means of spreading of the disease is seriously doubtful. Based on the navigation capabilities at that time and the incubation period of the disease, the crew of the ships should have already been wiped out on board before reaching Italy. The most likely scenario is that the disease spread to the Italian ports through other ships arriving from Greek ports (closer to Italy than Constantinople) which had already been infected by the disease. In any case, the belief that the disease reached the Italian ports through the Genoese galleys is also found in the 14th–century *Chronicon Senense* of Agnolo di Tura del Grasso.[607]

Medieval narrations are based on the contemporary understanding of infectious diseases, and their narrative plot is simply the result of collecting and recounting the most dramatic elements of individual accounts. Time is often contracted in these narrations, thus making it almost impossible to track the course of a specific disease. Nevertheless, it is certain that plague appeared in Messina and Catania in October 1347, and in Genoa and Venice in January 1348. It is troublesome to temporally locate the emergence of the disease in the Italian hinterland, although it is highly probable that the arrival of *Y. pestis* occurred from the north, rather than via Sicily.

According to de' Mussi and his narration of the great epidemic (*Istoria de morbo sive mortalitate que fuit de 1348*), the galleys that departed from Kaffa after the termination of the siege spread the disease throughout the Mediterranean. He mentions that [608]:

---

604 Wheelis 2002.
605 Derbes 1966.
606 Ayyadurai 2008; Eisen 2008.
607 Tura del Grasso, *Chronicon Senense*, 1347 (ed. Muratori 1729: 120.A11-C12).
608 Derbes 1966; Tononi 1884.

Nauigantes, cum ad terras aliquas accedebant, ac si maligni spiritus comitantes, mixtis homini-
bus interierunt, omnis civitas, omnis locus, omnis terra et habitatores eorum utriusque sexus,
morbid contagio pestifero uenenati, morte subita corruebant.

When the sailors arrived at those places and mingled with the locals, it was as if they had
brought evil spirits with them: every city, every settlement, every place was poisoned by the contagious disease and the residents, both men and women, died instantly.

By linking historical and epidemiological data and relying on rational reasoning, we
can complement—to some extent—the account of de' Mussi, and compensate for the
temporal discontinuities, but also identify the weaknesses of the narration. In 1345,
one year after his first unsuccessful attack on Kaffa, the Mongol Khan Jani Beg (d.
1357) returned for a second siege. During the second year of the siege (1346), an epidemic broke out in the ranks of the Mongol army. Jani Beg's wrath for his second failure coupled with his hatred for the Genoese led him to avenge Kaffa by catapulting
the shredded carcasses of plague victims in the city. Furthermore, in late 1346, the
decimated Mongol army put an end to the siege and retired. It is reasonable that
the Genoese informed—as required—their proveditor in Peran, Constantinople, of
the outcome of the war. The disease had already begun to take its toll on the residents of the city, but perhaps the authorities were expecting a brief epidemic outbreak. Obviously, subsequent developments contradicted their expectations, and
they started seriously thinking of evacuating Kaffa. The evacuation of the city
must have taken place one or two months after the end of the siege, possibly in February or March 1347, when the epidemic was at its peak. The only rational and feasible solution for them was to seek shelter in Peran, the metropolitan colony of
Genoa in Constantinople. It is worth noting that, despite the fact that the army
and part of its residents abandoned Kaffa, the Mongols made no effort to occupy
the infected city. The phrase of de' Mussi for those "who escaped Kaffa on the
ships" may indicate that not everyone was able to abandon the city, or that some
of its residents chose to resort to the countryside. The mention that some ships headed to Venice and others to Christian territories may mitigate, to some extent, the historical abomination of Genoa, since many Venetians lived in the besieged city and
they, too, decided to return to their homeland. But it also signifies the massive abandonment by the Christian population. Nevertheless, the fact that part of the population abandoned the city was perhaps not the result of panic and impulse, but rather
of an organized plan in collaboration with the Genoese residents of Peran, since no
commander would so recklessly dare to take a decision that would render defense-
less the most important colony of *Superba Genova* in the Black Sea.

The first mention of plague outside the Byzantine Empire concerns Messina and
Catania in October 1347. Messina is located 1,600 nautical miles from Kaffa. Any direct transmission of the disease from Kaffa to Sicily should be considered impossible.
Considering that plague broke out in Constantinople in the spring of 1347, it seems
more reasonable that the disease reached the ports of Sicily through the ships that

set sail from the ports of the Byzantine Empire, thus diminishing the importance of the nationality of those ships.

However, the report of the Byzantine historian Michael Panaretos (c. 1320-c. 1390) contradicts medieval and modern allegations of a direct spread of the disease to Italy through the controversial fugitive armada.[609] This information is directly related to the events in Kaffa, as it appears that, even after the end of the siege, the colony and the Genoese commercial station in Trebizond were in constant communication. de' Mussi's dramatic recount of the events leads to the conclusion that Kaffa was deserted, but this was not the case. The colony remained under Genoese control, the situation was normalized, and the naval base of Crimea became a foothold for the raids and lootings of the Genoese against Cherson (near Sevastopol in Crimea), apparently as retaliation for their friendly attitude towards the Mongols during the siege. As for the Trebizond epidemic, the outbreak of plague may have served as the pretext that Emperor Michael Megas Komnenos (c. 1288–1355) was seeking to move against the Genoese in the city. The epidemic outbreak coincided with the invasion of the Genoese station by a raging crowd, possibly under the guidance of Michael (with or without the encouragement of the Venetians residing in Trebizond). It is also possible that the invasion stemmed from the desperation caused by the epidemic and the consequent blame game for the spread of the epidemic among the Genoese ranks. The invasion resulted in the massacre of the merchants of Genoa, which in turn, lead to the decimation of Cerasus by the Genoese in 1348 as a retaliation to Michael.

The account of Panaretos is particularly interesting, as it mentions [610]:

... Μηνί Σεπτεμβρίῳ ,ϛωνς' ἰνδικτιῶνος α', ἐγένετο αἰφνίδιος θάνατος, ἡ πανούκλα, ὥστε ἀπεβάλλοντο πολλοὶ τέκνα καὶ συνεύνους, ἀδελφοὺς καὶ μητέρας καὶ συγγενεῖς, καὶ διεκράτησεν εἰς μῆνας ζ' ...

... in September of the year ,ϛωνς' of the first indiction, the plague caused sudden death and people lost their children, spouses, siblings, mothers and relatives, and this lasted for seven months ...

While the Romans calculated time since the legendary founding of the city of Rome (*Ab Urbe Condita* or *Anno Urbis Conditae*), which—according to the estimates of Dionysius Exiguus (470–544 A.D.)—corresponded to 753 B.C., for the Byzantines time began in 'the year after creation' (ἀπὸ κτίσεως κόσμου or *Ab anno mundi*).[611] According to the 70 Apostles, time began in 5509 B.C. Thus, based on the Byzantine calendar, the year ,ϛωνς' mentioned by Panaretos would have been the year 6856 after creation. To convert this date to the modern dating system, we need to subtract the year after creation (5509) from 6856, which gives us 1347. In his account, Kantakouzenos

609 Lampsidis 1958.
610 Lampsidis 1958.
611 Philip 1921: 41–58.

mentions that, in the spring of 'the same year ς'', which is an abbreviation for the year ͵ϛωνς' of Panaretos, a plague epidemic broke out in Constantinople. The epidemics of Constantinople and Trebizond seem to have taken place during the same Byzantine "calendar" year.

At this point, however, there is a critical detail that could challenge our understanding of the course of plague from Crimea to Constantinople. If the Byzantines employed the modern year numbering system, the epidemic of Trebizond would have taken place in September 1347, that is, several months after the epidemic of Constantinople. Among the legacies of the Roman World to the Byzantine Empire, was the *Julian calendar* (proposed by Julius Caesar [100 – 44 B.C.]). The Byzantine *Indiction*, a residue of the Roman indiction, was a peculiar fiscal cycle of 15 years which had no relation with astronomy, but was rather related to the imposition of taxes or the announcement of tax revisions (Latin *indictio*, meaning *announcement*).[612] Maybe the indiction is related to the 15-year periodic census of the Empire's tax revenue. Since the time of Constantine the Great, the indiction became a unit of time, also known as the *Constantine Indiction*, whereas, during Justinian's reign, it was introduced as the formal dating system of public documents and court records, starting on the first day of September. Thus, each year of the 15-year cycle began on the first day of September, starting from the first indiction of September 1, 312 A.D.[613] In the early fourth century, the first day of September was established as the beginning of the ecclesiastical year of the Greek-Orthodox Church, which coincided with the beginning of the indiction.[614] This day is considered as the Byzantine New Year's Eve, and a ceremonial liturgical procession took place, known as the Ἀρχὴ τῆς Ἰνδίκτου (*Beginning of Indiction*, i.e. the beginning of the Ecclesiastical New Year); even in the contemporary calendar of the Orthodox Church, the first day of September is marked as the 'beginning of the indiction'. Apparently, the difference lies in the different points of time within the same year now (1 January) and then (1 September). The confusion caused by the different medieval calendars becomes more apparent by considering that, when the spring outbreak occurred in Constantinople, the Byzantine year was already halfway (August–September 1347) while Venice and Genoa were still going through 1346, celebrating New Year's Eve when the Byzantine calendar showed March 1, 1347, and March 25, 1347 respectively.

In an attempt to imitate the Byzantine way of thinking, the events that took place in Constantinople in the spring of Kantakouzenos' 'same year ς'' correspond to the year ͵ϛωνς' of Panaretos (6856 or 1347), which had already begun on September 1, according to the Byzantine calendar. This means that the first major city that was affected by plague through Kaffa was Trebizond and not Constantinople. When we translate Byzantine dates according to the modern system of dating, however, it

---

612 Philip 1921: 41–58.
613 Philip 1921: 41–58.
614 Philip 1921: 41–58.

comes as a surprise that the epidemic of Trebizond followed that of Constantinople. The lack of sources regarding the exact time when the siege of Kaffa ended leaves one more possibility open. Since the Byzantine year stretches from September to August, September belongs to the previous year based on the modern calendar. The calculation (6856 – 5509 = 1347) refers to the year in general. The reference of Panaretos to "September of the year ,ϛωνϛ′ (1347) of the first indiction" automatically removes another six months from the modern system of dating, taking us to autumn of 1346, when the siege of Kaffa was coming to its end. The other reference of Panaretos to the sudden death that "lasted until the month ζ′" corresponds to March when we begin our calculation from September 1, namely the spring period mentioned by Kantakouzenos. Therefore, the possibility that the disease reached Constantinople through Trebizond seems to be an alternative scenario.

The reference to the October epidemic of Messina comes from the chronicler Michele de Piazza (d. 1377). It was made several years after the first appearance of the *Black Death*, during the writing of the *History of Sicily* (*Historia Sicula*). In fact, Michele de Piazza mentions that his information stems from the narration of a Franciscan monk, who had written his own account of the events ten years earlier pointing out that twelve Genoese galleys traveling from the East had brought plague to Messina in October 1347, although Kaffa was not mentioned in his account. Obviously, it was easy for each writer to reproduce these views, which Gabriele de' Mussi presented in such a convincing manner that even today the siege of Kaffa is associated with the spread of plague to Europe.

de' Mussi's text seems more valuable as the chronicle of a siege rather than as a source of epidemiological interest. The collation of facts is rife with time discontinuities and ambiguities, but, in any case, it retains its historical value. It seems that plague swept the Greek coasts and Venetian-held islands in the Aegean first, before reaching southern Italy, and then spreading throughout Europe. Before the disease appeared in southern Italy, it had already affected Euboea, Crete, Lemnos, Thessaloniki, Methone, and Koroni.

Although it could all have begun in Kaffa in 1346, the course of *Y. pestis* towards the West has to be dissociated from the narration of de' Mussi. After the end of the siege, the Genoese may have transmitted the disease to Trebizond, wherefrom it reached Constantinople. The epidemic of Constantinople in 1347 and the ships that departed from the capital and other Greek ports, however, seem to have been the tide that swept Europe.

Pax Tatarica and the Origins of the *Black Death*

The sudden reemergence of the *Black Death* has been the subject of extensive medical and historical research, focusing on the identification of the origins of plague in the fourteenth century. The three strains of the disease known to date, namely *Yersinia pestis* Antiqua, *Y. pestis* Medievalis and *Y. pestis* Orientalis, coincide with the three respective pandemics, the *Plague of Justinian*, the *Black Death*, and the *Plague of the late 19th century*.[615]

Based on the prevailing view, plague became endemic in at least two areas of the planet after the end of the first pandemic, that is, in Africa and Asia. It gradually moved in other directions and affected new populations which had never contracted the disease before. Most scientists support the theory that plague originated in Central Asia, wherefrom the Mongols are believed to have spread it to the European continent.[616] The role of the Mongols was catalytic, not so much because of the events in Kaffa, but rather due to the stability that resulted from the unification of Eurasian tribes by Genghis Khan (reign. 1206–1227) during the thirteenth century. Until then, the tribes of the steppe had been engaged in perpetual wars that led to the obstruction of trade relations with southeastern Europe and the Arabic world. The vast, yet united Empire of Genghis Khan secured a coveted regularity that allowed caravans to safely travel to the East, which seems to have made it easier for *Y. pestis* to spread—slowly but steadily—initially to the East and then to southeastern Russia.[617]

At the height of its power, the Mongol Empire (1280–1350) succeeded in uniting a vast territory of major geostrategic and commercial importance that included the whole of China, most of the Russian Far East, Central Asia, Iran, and Iraq. Christian leaders were relieved by the Mongol presence in the Middle East because, in addition to their trade agreements, Islam was now facing a deadly enemy.

The unification of Eurasia under Mongol control led to the *Pax Tatarica*, and the administrative organisation of the conquered areas into *uluses* created an ideal environment for previously isolated populations to come into contact with each other. The Khanate of the Golden Horde that controlled the area extending from the Lower Danube River to Central Asia proved condescending to European economic penetration and to developing relations with the West in general, until its relationship with Venice and Genoa reached extremes. The political stability of the Mongolian power and the flourishing trade between Asia and Europe served as the fundamental conditions for the gradual introduction of plague to virgin populations.

Of great medico-historical significance is the account of Arabic writer and chronicler Ibn al Wardi (1292–1349 A.D.). He recounts the testimonies of Muslim merchants who returned from Syria and Crimea, where they had witnessed an epidemic outbreak in the land of the Golden Horde in October–November 1346. In line with the general perception of Arabs who saw the Mongols as barbarians, Ibn al-Wardi mentions that the epidemic in the territories of the Golden Horde started from the 'land of darkness', an area that some researchers identified with Mongolia.[618] A second element that many researchers consider supportive to the theory of Asian origins are the

615 Guiyoule 1994.
616 Norris 1977.
617 Norris 1977.
618 Norris 1977.

archaeological findings in the Nestorian necropolises near the lake Issik Kul, in modern Kyrgyzstan, where mortality rates seem to have increased between 1338 and 1339, while the tombstones are inscribed with references to a virulent disease that decimated the population. More specifically, among the 330 tombstone inscriptions that were found and cover the period 1186–1349, 106 are related to the period 1338–1339, that is, seven years prior to the events in Kaffa.[619]

Besides Ibn al Wardi, two other Arabic scholars speak of the epidemic of the 'land of darkness': Ibn Khatib (1313–1374 A.D.) and Ibn Khatimah (d. 1369). Their accounts have been the object of great controversy due to the mention that the disease started in the 'land of Khitai'.[620] Ibn Khatib and Ibn Khatimah may have referred to the kingdom of Khitai, who was the king of the short-lived Turkic-Utigur kingdom of Kara Khitai located south of the Caspian Sea.

China is also considered as a candidate country of origin of the *Second Pandemic*. Existing information on China's epidemics stems from the *Imperial Encyclopedia* (*Ku Chin T'u Shu Chi Ch'eng*) of 1726, which records the diseases and natural disasters that affected the Empire.[621] According to *Ku Chin T'u Shu Chi Ch'eng*, there are various reports of epidemics with high mortality rates. Prior to 1346, 26,000 losses were reported in the province of Chekiang in 1308, along with the epidemic outbreaks in Hubei in 1313, 1320, 1321, 1323, and 1331. In fact, it has been reported that, during the latter epidemic, the region lost nine tenths of its population. In 1345, an epidemic broke out in Fukien and Shantung that lasted until 1346. During the *Second Pandemic* in 1351, an epidemic struck the provinces of Shansi, Hubei, and Kiangsi, with a mortality rate that reached 50%.[622]

According to another theory, plague moved northwards from the Great Lakes of Central Africa to Central Asia, where it became endemic among the marmot populations, and then moved towards West and East.[623] As far as the role of rodents is concerned, it has been suggested that plague was enzootic among the wild rodent populations in Kurdistan and North Iraq, wherefrom it spread to Southeast Russia over the next centuries. From there it reached Europe and the Middle East, Turkestan, the steppes of South Siberia, and probably Mongolia, but definitely not China, which is believed to have been infected through India and Burma.[624]

According to Kantakouzenos and Gregoras, South Russia was at the epicenter of the epidemic. More specifically, Gregoras reports that the terrible "*thanatikon* started from the Scythians at the outfall of the River Don" ("ἀρξάμενον ἀπὸ Σκυθῶν καὶ Μαιώτιδος καὶ τῶν τοῦ Τανάϊδος ἐκβολῶν"). Kantakouzenos also vaguely reports that the disease originated from the Scythians of the Hyperborea region (that is,

---

619 Norris 1977.
620 Norris 1977.
621 Norris 1977.
622 McNeill 1976: 263.
623 Dolls 1977: 13–67.
624 Norris 1977.

the people living 'beyond the North'): "the *thanatikon* started from the Hyperboreans and originated from the Scythians" ("ἐκ τῶν Ὑπερβορέων πρῶτον ἀρξάμενος Σκυθῶν"). We must not forget, however, that contemporary people had imprecise geographic knowledge, and thus defined places and peoples with a single word. The concept of the *Hyperboreans* originated in Ancient Greece, and referred to a vast area entailing all peoples living north of the Balkan Peninsula as well as the Scythians of the Black Sea. This indiscriminate ethnographic grouping of peoples was a common phenomenon in the Middle Ages, not excluding the Byzantine Empire.

Another realistic possibility is that the inland of Asia Minor had already been affected by plague prior to the events in Kaffa. In May 1345, the dauphin to the throne of France, Humbert II of Viennois (1312–1355), led a Papal coalition with crusaders from Venice, Knights Hospitallers of Rhodes, and the Kingdom of Cyprus, against the Turkish Emirate of Aydin. There is reason to believe that, by the time Humbert departed from Smyrna in 1346, his army had suffered heavy losses during its campaign against the Ottomans in inner Asia Minor due to an epidemic, possibly of plague.[625]

If this was the case, it is possible that plague had already started to invade inner Asia Minor from the depths of the East at the time of the Mongols' siege of Kaffa. Theoretically, this means that the disease had already settled in the territories of the Emirates of the Turkish Ghazi warriors or the Mongol Ilkhanate that had expanded as far as Asia Minor by that time.

This event can be supplemented by Panaretos' account of the Trebizond epidemic, thus illustrating the numerous alternative courses of plague towards the capital. If the disease had been limited to Crimea, the significance and value of accounts and reports on the transmission of the disease to Constantinople by the Genoese could have been strengthened. By taking the report of Michael Panaretos into consideration, however, we should not exclude the possibility that plague had followed a different course to Constantinople rather than through Crimea. The report of Panaretos on the Trebizond epidemic of 1347 is not the first of its kind, as another epidemic of sudden death had been reported in 1341 in Pontus. The medieval sources often employ the words '*sudden death*' to describe epidemic outbreaks of plague. The possibility that plague emerged periodically in Asia Minor prior to 1346 can neither be proven with accuracy nor excluded with certainty.

It is understood that historical sources require delicate interpretation. de' Mussi's account has been misleading. Therefore the place and time of the reemergence of plague may have to be redefined. After all, the inaccurate dating of events or the ignorance of differentiated sources can cause problems, since they disorient scientific research from the right geographic areas that played a key role in the spread of plague.

Fundamentally, two questions need to be answered, namely *where* Y. pestis appeared first, and what was the place of origin of the disease. By summing up all

---

625 Jorga 1896: 69–71; Michaud 1852: 111–112.

available historical data, it is possible to redefine the course of the disease. In line with the view expressed in the previous section on the basis of historical evidence concerning the possible existence of natural enzootic foci in the areas of Syria and Mesopotamia, the gradual spread of *Y. pestis* from the Middle East to the East could be possible. The course of the disease towards the East should be investigated in light of the impact of the conflicts between the Arabs and the Mongols during the thirteenth century. The Mongol and the Muslim worlds came into contact in 1251, when the former conquered the Abbasid Caliphate and founded the Ilkhanate, which was subordinate to the Mongolian Empire and became a permanent threat to the Middle East, while the Euphrates River was the only barrier that contained the vehemence of the steppe warriors.[626] The hordes of Hulagu Khan (ca. 1218– 1265 A.D.) captured Baghdad in 1258 and Damascus in 1260, along with its surrounding area located within the boundaries of the aforementioned natural focus of the disease. The ensuing peace and normalization of the political situation in the Middle East allowed caravans to use the old commercial routes once again, and opened new ways to the depths of the East.[627] Plague may have gradually spread north towards the Caspian Sea and the Caucasus. Furthermore, some researchers suggest that fur trade may have contributed to the more rapid spread of the disease, since fleas remained attached to the furs.[628] The most plausible scenario, however, suggests that the rodent populations of Caucasia slowly spread the disease towards the East, with the marmot and gerbil species becoming enzootic in these geographic regions.

Regardless of when it appeared, plague is subject to specific epidemiological patterns just like any other epidemic. An epidemic has to be characterized by close interactions between susceptible vectors and reservoirs. The infectivity of *Y. pestis* is determined by the number of microbes, the susceptibility of fleas to infection, the ability of fleas to infect animals and people, and the susceptibility of animals and humans. Apparently, environmental conditions played a crucial role in the phylogenetic evolution of *Y. pestis*, which was able to survive in the new environment, while plague gradually became enzootic in wild rodents in the Caucasian region of Russia prior to the *Black Death*.

Studies thus far have shown that *Y. pestis* is a phylogenetic clone of *Y. pseudo-tuberculosis*.[629] The molecular analysis of 70 strains that belong to the three known biotypes revealed the existence of two ribotypes, O and B, in 65.7 % of the strains studied (ribotype is the RNA component of the genetic material). Ribotype B is found in the strains of the biotype *Y. pestis* Orientalis, while ribotype O is found in *Y. pestis* Antiqua and *Y. pestis* Medievalis.[630] This has helped to geographically clas-

---

626 Curtin 1908: 22– 247.
627 Norris 1977.
628 Ampel 1991.
629 Whittam 1996; Achtman 1999; Paradis 2005.
630 Drancourt 2002.

sify them, and ribotypes B can currently be found in all continents, while ribotype O is located exclusively in Central Africa and Central Asia.[631] The biotypes of *Y. pestis* that were found in the regions of Transcaucasia and Central Asia can be classified according to the ability of bacteria to ferment glycerol and arabinose, and to the reduction of nitrate to nitrite.[632] The strains of the Caucasus region are a special group that is phylogenetically considered as one of the most ancient strains of *Y. pestis*; they are identified as *pestoides*.[633] The pestoides strains can be found in areas of the former Soviet Union (Northern Caucasus, Dagestan, Altai and Talas mountains).[634] These strains are characterized by a loss of the small 9.5-kb plasmid (pPCP1) and ineffective encoding of bacteriocins (proteins produced by bacteria to inhibit the growth of other species of bacteria).[635] They parasitize specific rodent populations and present a series of metabolic anomalies in the glycerol cycle.[636]

The classification method of *Y. pestis* strains based on the fermentation of glycerol and nitrate reduction to nitrite is useful for identifying particular characteristics of their place of origin.[637] Based on existing findings, *Y. pestis* Antiqua (fermentation of glycerol and reduction of nitrate to nitrite) has been found in Central Asia, whereas *Y. pestis* Medievalis (fermentation of glycerol without reduction of nitrate) in the areas of Southeast European Russia. The ability of glycerol fermentation seems to have allowed *Y. pestis* to survive in the wider region of Southeast European Russia, while the microbe's capability to persist in specific reservoirs may be due to this specific ability.

It is believed that the ability of the *Y. pestis* Medievalis strain to ferment glycerol is directly related to the need of an ideal reservoir during the period of animals' winter hibernation.[638] *Y. pestis* has obviously managed to survive in the harsh environmental conditions of the area by taking advantage of two distinct species of terrestrial rodents, marmots and gerbils. Marmots (*Marmota spp.*) become infected with *Y. pestis* before the start of their winter hibernation.[639] Prior to their hibernation, animals ensure that they have stored large quantities of fat. Throughout hibernation, the organism undergoes periods of full inactivity and periodic awakenings that do not fully awaken the animal. These phases alternate and—depending on the species—last from a few days to five weeks, until the final awakening of the animal.[640] During hibernation, all normal functions are suppressed to such an extent that even

---

631 Drancourt 2002.
632 Zhou 2004.
633 Bobrov 1997; Garcia 2007.
634 Motin 2002.
635 Pourcel 2004.
636 Motin 2002.
637 Martinevsky 1973.
638 Carey 2003.
639 Carey 2003.
640 Burlington and Klain 1967: 701–708.

the respiratory function decreases from 100 to 200 breaths per minute to four to six breaths per minute, with periods of apnea being observed in some species.[641] During hibernation, the survival of the organism depends on hydrolysis of the stored fat (conversion of fat in glycerol and fatty acids).[642] The glycerol released from adipose tissue hydrolysis is used as a substrate by the liver and kidneys for the production of glucose through gluconeogenesis.[643] When *Y. pestis* invades the organism, it binds and entraps glycerol in the bacterial cell and its fermentation ensues.[644] The gradual depletion of glycerol stocks by *Y. pestis* forces the animal organism to produce more glycerol, as its molecule protects tissues from freezing. [645] This allows *Y. pestis* to ferment glycerol throughout hibernation.

Besides the role of marmots, great gerbils (*Rhombomys opimus*), which are the main hosts of *Y. pestis* in the steppes of Central Asia, may have been an important epidemiological factor in the survival of the bacterium and the onset of plague outbreaks and epizootic.[646] Gerbils live in Caucasia and Central Asia and are a rodent species of great interest in the sense that they move in groups over long distances of up to three miles (over five kilometers), while they are able to survive even at temperatures of −20°C without the need for hibernation.[647] The population of gerbils in a specific area can sharply increase as a result of environmental conditions, since whole groups of rodents migrate in order to join other groups, thus demonstrating an unprecedented social behavior. These movements are nowadays the subject of epidemiological study and surveillance, because of the emergence of epizootic among the ranks of this species and the development of natural foci in Central Asia.[648] In fact, it is speculated that the drastic climatologic and environmental changes that take place today are bound to drive gerbils into great movements, thus leading to new natural foci and epizootic outbreaks in Asia.[649] The antibodies of gerbils and their dynamic immunity after becoming infected with *Y. pestis* present particular fluctuations. It is remarkable that gerbils infected during summer periods survive, whereas infection during winter becomes fatal after spring.[650] This dynamic seasonal fluctuation of plague immunity may play an important role in the preservation of *Y. pestis* in host rodent populations.[651]

Therefore, it should not be considered impossible for *Y. pestis* to survive through susceptible rodent populations like marmots and gerbils, thus making the disease

---

641 Carey 2003.
642 Burlington 1967.
643 Burlington 1967; Green 1984.
644 Willet 1992: 57.
645 Willet 1992: 57.
646 Kausrud 2007.
647 Kausrud 2007; Randall 2002.
648 Kausrud 2007.
649 Kausrud 2007.
650 Park 2007.
651 Begon 2006.

enzootic in the areas of the Caucasus, long before the first records of the reemergence of plague during the fourteenth century. Taking the historical changes brought about by the unification of Eurasia by the Mongols into consideration, the disease may have first spread from the Caucasus to Central Asia, wherefrom it moved west again. Of course, the term 'land of darkness' by which the Arabic chroniclers liked to refer to the Mongol territories, is a very broad geographic concept. In this sense, it could well refer to any area of the vast Mongol Empire, although it was most likely related to the neighboring lands of the Arabic caliphates controlled by the Golden Horde of the Caucasus.

The Nature of the *Second Pandemic:* Bubonic or Pneumonic Plague?

Over the years, several theories have been proposed in numerous studies on plague epidemics. Some theories are related to the place of origin of the microbe, while others focus on the type of the disease, or explore alternative ways of disease transmission. Other alternative theories have been put forward at times, such as the implication of the Nosophyllus fasciatus and Pulex irritans fleas in transmission.[652] Other researchers have suggested that pandemics are not associated with Y. pestis, but rather with the disease caused by anthrax or some filovirus (Ebola-like or Marburg-like viruses).[653] Moreover, the type of plague (bubonic or pneumonic) has frequently been brought into question.

The accounts of Byzantine writers reveal the existence of two distinct forms of plague. Although the Byzantines adopted the concept of disease transmissibility through the miasma, they ultimately presumed that the two clinical forms of the disease were transmitted between humans. It soon became evident, however, that no direct contact with a patient was required for a human to become infected and ill with plague. Nevertheless, the view that the disease was transmitted from human to human through respiration did not cease to exist. This may be true when we take into account the way of transmission of pneumonic plague, but this medieval perception was rather part of the overall concept of contaminated air that prevailed at the time. The view that was put forward based on the epidemiological observations of French physicians regarding the time of death of whole families became very popular in the medieval world. More specifically, two scenarios were proposed. The members of the same family could pass away within 24 hours or within a period of three weeks. Physicians suggested that whoever had not become ill within three weeks was no longer at risk. Medieval societies accepted this theory that probably

---

652 Benedictow 2016: 1–4.
653 Benedictow 2010: 489–490.

served as a slight psychological help that offered hope to contemporary people, although it was not scientifically founded.[654]

Beyond any linguistic similarities or the influence of Thucydides' work, Kantakouzenos vividly describes the two forms of the same disease, which correspond to bubonic and pneumonic plague. As already mentioned, the clinical picture described by the Byzantines is sometimes fully identified with the known symptomatology and sometimes not, such as the accounts of Procopius and Kantakouzenos regarding the effect of the plague toxin on the central nervous system of patients, which causes coma, delirium, or manic excitation.

The narrative of Kantakouzenos has led many researchers to the diagnosis of pneumonic plague.[655] According to Kantakouzenos, if the disease reached the head, the patient would fall into lethargy and death would soon ensue. The infection of the lungs was accompanied by strong pains in the chest, haemoptysis, and intense thirst, while abscesses with black pus outflow appeared on the arms and jaws. Kantakouzenos suggested that the evolution of the disease was feasible if the abscesses were ruptured:

> ... σφάκελλος ἐξέρρει δυσώδης καὶ πολύς καὶ τό νόσημα διεφορεῖτο εἰς ἐκεῖνο, τὴν διενοχλοῦσαν ὕλην ἀπορρίπτον ...

> ... an abundant and malodorous pus flows and the development of the disease differs when this (the abscess) ruptures and releases its harmful matter ...

Kantakouzenos also refers to skin manifestations that probably correspond to petechiae spots on the skin.[656]

The hypothesis of pneumonic plague has been reinforced by the symptoms described, especially high mortality and the rate at which the disease spread. At the time of the *Black Death*, mortality due to bubonic plague ranged from 40 to 60 % of the population, while, for pneumonic plague, the rate reached 95 %.[657] We must not forget, however, the fact that medieval plague epidemics were primarily urban epidemics, in contrast to the modern outbreaks of the disease that affect remote rural areas. This explains the high rates of mortality.

Just like the first pandemic, the *Second Pandemic* spread at an amazing speed. The Black Death spread on a daily basis (*per diem*), compared to modern epizootic outbreaks that show an annual spread (*per annum*). It is speculated that the *Black Death* moved at a speed of 1.5 to 6 kilometers per day, namely at a rate much faster than the epidemics of the nineteenth and twentieth century.[658] When we compare the bubonic and pneumonic plague epidemics of the first and *Second Pandemic* with the

654 Ziegler 2003: 4–5.
655 Joannes Cantacuzenus, *Historiae*, IV.8 [A.C. 1347] P. 730 (ed. Schopen 1832: 50.16–21).
656 Joannes Cantacuzenus, *Historiae*, IV.8 [A.C. 1347] P. 731 (ed. Schopen 1832: 51.4–5).
657 Cohn 2008.
658 Cohn 2008.

respective epidemics of the third pandemic, we can find similarities in their symptomatology.

As already mentioned, the pneumonic form of plague can either be primary or secondary. The secondary form occurs in 5% of bubonic plague patients, who are considered to be the main cause of primary pneumonic plague, as they transmit the disease through droplets.[659] The pneumonic form has an incubation period of one to three days, and manifests itself through high fever, great malaise, intense coughing, dyspnea, and cyanosis. In secondary plague, these symptoms are accompanied by the symptoms of the glandular form. The prognosis is poor, and, if left untreated, the patient dies within 1–2 or 5 days.

The accounts of Byzantine and Western chroniclers reveal that the Byzantine Empire was faced with pneumonic plague, while Western Europe was affected by bubonic plague. To what extent, however, does this finding correspond to reality? In addition to the detailed account of Kantakouzenos, most sources speak of "the death of the buboes", and the only thing reminiscent of pneumonic plague is the gloomy description of sudden deaths. Based on modern epidemiological and microbiological data, the pneumonic form of plague may not be primarily manifested, but rather as a result of bubonic plague after *Y. pestis* spreads to the lungs.[660]

The resulting pneumonia has to be differentiated from the Acute Respiratory Distress Syndrome (ARDS), which is prevalent in the bubonic and septicemic form. However, in both forms of pneumonic plague the mortality rate is extremely high. Pneumonic plague is generally considered as an unusual form of plague, and it is the only form transmitted directly from human to human.[661] Constantinople may have been struck by such an epidemic in 1347, although this is considered improbable as previously said. However, the case of the hecatomb of Manchuria in 1910–1911 leaves open the possibility of primary pneumonic plague.[662] The information derived from other Byzantine and Venetian sources covers the spectrum of bubonic plague, which might eventually mean that Constantinople had the great misfortune of both clinical forms to coexist, a common phenomenon in cases of epidemics.[663] It can reasonably be inferred that the epidemic took the form of bubonic plague. The victims affected by secondary pneumonic plague transmitted the disease, apparently through droplets, to their relatives, who became candidates for the direct transmission of the disease and the onset of primary pneumonic plague.[664] Kantakouzenos' report on the symptoms may indicate the gradual predominance of the pneumonic over the bubonic form of the disease. Even in this case, however, research on other pneumonic plague epidemics of the twentieth century has shown that these

---

659 Poland 1999: 43–46.
660 Smith 1996.
661 Smith 1996.
662 Smith 1996.
663 Centers for Disease Control and Prevention 2001; World Health Organization 1968.
664 Pechous 2016.

were strictly local and self-limited epidemics with no dispersion.[665] We can therefore assume that something similar must have happened in Constantinople in 1347.

Finally, another interesting approach to the nature of the *Second Pandemic* was based on the immunity induced by plague. It has been suggested that medieval plague induced longer-lasting immunity compared to the brief immunity induced by modern epidemics, although this cannot be determined with certainty.[666] But the finding that long-term immunity can result from the organism's host cells binding to *Y. pestis*—which continues to survive in these cells—leaves open the aforementioned possibility.[667] This does not apply, however, to the cases of pneumonic plague in the pre-antibiotic era, given the nearly 100% mortality rate of patients.

It should be pointed out that the anatomopathological diagnosis of pneumonic plague in patients of modern epidemics is a rather easy process, although the lesions caused by pneumonia are not specific to plague.[668] Instead, the diagnosis of pneumonic plague in patients of the *Black Death* era is practically impossible.

Aggregations of Epidemics during the *Black Death* in the Byzantine Space

By grouping the epidemics that broke out in the Byzantine Empire during the fourteenth and fifteenth centuries, we are able to identify aggregations of epidemics and isolated outbreaks similar to the first pandemic. A more accurate dating system could help in classifying plague outbreaks into major and minor aggregations that affected the Empire. In the 107 years between the first outbreak and the fall of Constantinople, 54 years were characterized by plague epidemics in the Southeast Mediterranean, and 53 years were free from the disease. A total of nine major aggregations of epidemics can be distinguished, along with 16 local outbreaks.

The geographic distribution of epidemics shows that Constantinople and its surrounding region and the Peloponnese were the areas most affected by plague, followed by Crete, Cyprus, and the Ionian islands. This distribution is presented in more detail in Table 5.

Based on existing sources, Constantinople appears to have been the city most often affected by the disease, approximately every 11.1 years. Methone and Koroni, two cities in the Peloponnese under Venetian control, constitute an interesting case, since they were both affected simultaneously in all four epidemics. This finding of 'ping-pong' contamination should not come as a surprise, given that these cities served as major Venetian naval bases, and communication between them was regu-

---

665 Lien-Teh 1922; Seal 1969.
666 Walløe 2008.
667 Titball 2001.
668 Aufderheide 1998: 195–198.

Table 5. Epidemics per geographical area during the *Black Death:*

Areas	14th–15th c. (1347–1453)	Total epidemics and duration in years
Constantinople (Thrace)	1347, 1361–1364, 1379–1380, 1386, 1391, 1397, 1403, 1409–1410, 1421–1422, 1431, 1435, 1438, 1441, 1448	14 (20)
Macedonia	1347, 1372, 1378, 1422–1423	4 (5)
Central Greece[1]	1388, 1423, 1426, 1448	4 (4)
Epirus	1368, 1374–1375	2 (3)
Peloponnese[2]	1347–1348, 1362–1363, 1375, 1381–1382, 1390–1391, 1397–1402, 1410–1413, 1418, 1422, 1431, 1441, 1448	12 (23)
Islands of Aegean Sea	1347–1348, 1362, 1408, 1445	4 (5)
Ionian Islands (Venetians)	1400, 1410–1413, 1416, 1420, 1450	5 (8)
Crete (Venetians)	1347, 1362–1365, 1375, 1388–1389, 1408–1411, 1418–1419	6 (14)
Cyprus[3]	1348, 1393, 1409–1411, 1419–1420, 1422, 1438	6 (9)
Pontus	1347, 1362–1363, 1435	3 (4)
		60

[1] Venetian and Byzantine possessions.
[2] Venetian and Byzantine possessions, Principality of Achaea.
[3] French House of Lusignan & Venetian possession.

lar (the distance between them is only 16.7 nautical miles). Finally, Thessaloniki and Trebizond were next in order with three epidemics.

The organized maritime network of the Aegean contributed to the spread of plague to the islands. Crete and Cyprus were severely affected due to their strategic position. The epidemic that broke out in Venetian Corfu in 1410 is obviously related to the epidemic that struck Methone and Koroni in the same year, which was spread throughout the Ionian Sea by the Venetian galleys. A similar case is the epidemic that broke out in Karpathos in 1408, which was apparently related to the epidemic in neighboring Crete.

The first aggregation of the *Second Pandemic* in Byzantium occurred between 1347–1348 (Map 10). It affected Constantinople, Euboea (Negroponte), Crete, Lemnos, Thessaloniki, Trebizond, Methone, and Koroni, as well as the inland of the Peloponnese, Rhodes, and Cyprus.

The sources make no reference to plague over the next twelve years, until the next aggregation that occurred between 1361 and 1365 (Map 11), which more or

Table 6. Epidemics of *Black Death* in the 'Byzantine Space' of the Eastern Mediterranean (14th–15th c.).

Cities	14th century	15th century	Total
Constantinople	1347, 1361, 1364, 1386, 1391	1409, 1421, 1431, 1435, 1438	10
Adrianople (Edirne)	1362		1
Athens	1388	1423	2
Thessaloniki	1347, 1372	1423	3
Ioannina	1368, 1375		2
Gallipoli		1403	1
Koroni	1347, 1399	1402, 1410	4
Nicosia		1438	1
Methone	1347, 1399	1402, 1410	4
Patras		1431	1
Peran	1379	1441	2
Selymbria		1448	1
Trebisond	1347, 1362	1435	3
Negroponte (Halkida)		1426, 1448	2
Candia (Heraklion)	1397		1

Table 7. Epidemics in the Islands of the Ionian Sea and the Aegean Archipelago (1347–1453).

Islands (under Byzantine or Venetian rule)	14th–15th c.	Total
Karpathos	1408	1
Corfu	1400, 1410–1413, 1420, 1450	4
Cephalonia	1416	1
Cyprus	1348, 1393, 1409–1411, 1419–1420, 1422, 1438	6
Crete	1347, 1362–1365, 1375, 1388–1389, 1408–1411, 1418–1419	6
Lemnos	1347, 1362	2
Rhodes	1348	1
Chios	1445	1

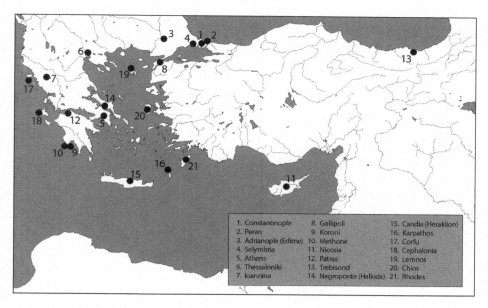

1. Constantinople	8. Gallipoli	15. Candia (Heraklion)
2. Peran	9. Koroni	16. Karpathos
3. Adrianople (Edirne)	10. Methone	17. Corfu
4. Selymbria	11. Nicosia	18. Cephalonia
5. Athens	12. Patras	19. Lemnos
6. Thessaloniki	13. Trebisond	20. Chios
7. Ioannina	14. Negroponte (Halkida)	21. Rhodes

Map 9. Cities affected during the *Second Pandemic* in Byzantium (1347–1453).

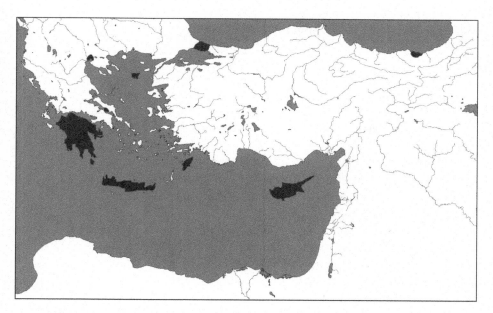

Map 10. Aggregation of epidemics during the period 1347–1348.

less affected the same regions. Starting from Constantinople in 1361, plague gradually spread to Adrianople, Trebizond, Lemnos, Crete, Cyprus, and the Peloponnese. In fact, judging from the account of Panaretos regarding the Trebizond epidemic

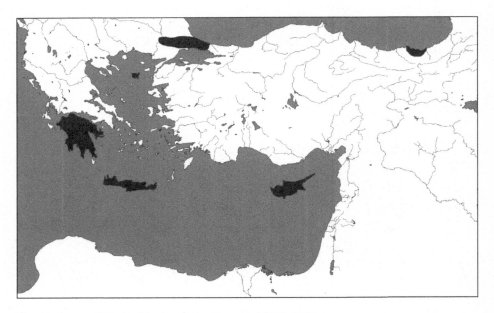

Map 11. Aggregation of epidemics during the period 1361–1365.

of 1362, when 'sudden death befell on the Turks' (ˌςωνςˈ ἐπεισπεσόντα τοῖς Τούρκοις αἰφνίδιον θάνατον), it appears that the epidemic had spread to the inland of Asia Minor.[669]

The third aggregation seems to have originated in Thessaloniki in 1372. By 1376 it had spread to Epirus, the Peloponnese, and Crete (Map 12). The next references indicate that the fourth aggregation took place between 1378 and 1382, and affected Mount Athos, Galatas, Genoese Peran, and the Peloponnese (Map 13).

In 1386, a fifth aggregation occurred, and lasted until 1391 (Map 14), starting from Constantinople and moving to Athens, the Peloponnese, and Crete. Constantinople was affected again during the sixth aggregation that took place between 1397 and 1402, but plague mainly swept the Venetian possessions in Candia, Koroni, Methone, and Corfu (Map 15).

The seventh aggregation, which was one of the most severe, occurred in the period 1408–1413 (Map 16). It mainly struck the islands and the capital. Once again, the Venetian possessions were harshly affected, and apparently the disease was spread from one to another, with Crete, Cyprus, Koroni, Methone, and Corfu being its victims. The eighth aggregation took place between 1417 to 1423 (Map 17) and struck Constantinople and almost all of Greece. This epidemic spread to Macedonia, Epirus, Central Greece, and the Peloponnese. The Venetians did not remain unaffected since

---

669 Lampsidis 1958.

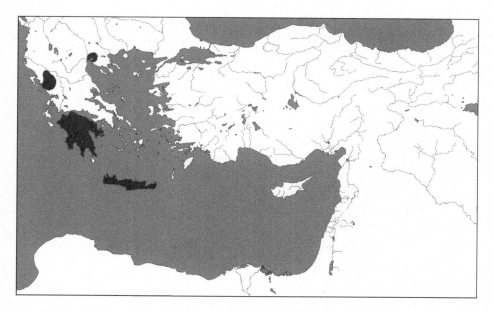

Map 12. Aggregation of epidemics during the period 1372–1376.

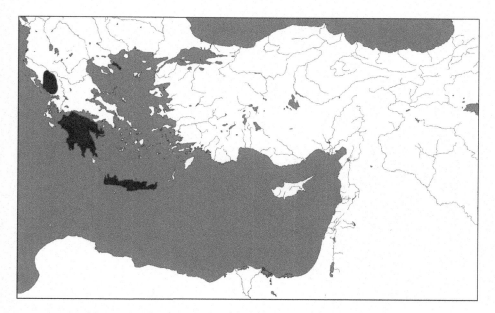

Map 13. Aggregation of epidemics during the period 1378–1382.

Cyprus, Crete, and Corfu were infected once again. The ninth and last aggregation broke out in 1435 in Constantinople, and the epidemic reached Trebizond.

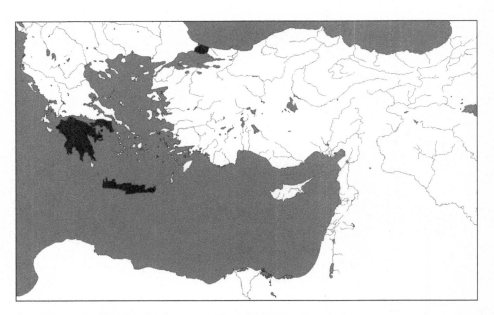

Map 14. Aggregation of epidemics during the period 1386–1391.

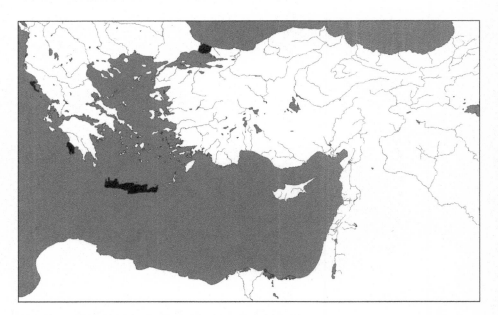

Map 15. Aggregation of epidemics during the period 1397–1402.

At the local level, epidemic outbreaks occurred in Ioannina (1368), Cyprus (1393), Gallipoli (1403), Cephalonia (1416), and Chalkida (1426). A series of outbreaks occurred in various areas of the Byzantine region during the decade of 1430 and until 1453,

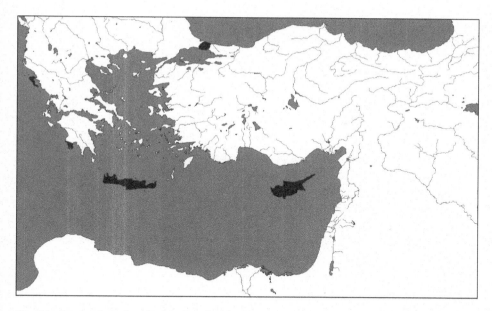

Map 16. Aggregation of epidemics during the period 1408–1413.

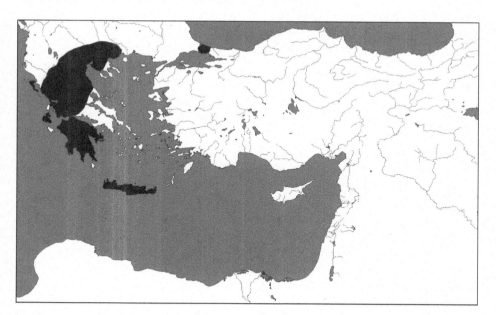

Map 17. Aggregation of epidemics during the period 1417–1423.

which may have taken place in the same year, but affected areas that were remote from each other. These outbreaks occurred in Patras and Constantinople in 1431, Nicosia and the outskirts of Constantinople in 1438, Peran and the Peloponnese in

1441, and Selymbria of Thrace, the Peloponnese, and Negroponte in 1448. These three cases could theoretically be associated to each other, but the ambiguity of our sources does not allow us to group them into an aggregation. Finally, two out of the sixteen outbreaks of this period were recorded in Chios (1445) and Corfu (1450).

It is difficult to compare the periodicity of the two pandemics due to the different number of years under consideration, keeping in mind that the scope of the present study does not extend beyond 1453. Nevertheless, we are able to identify the general trends of the epidemics in these two periods. During the 208 years of the first pandemic, 79 epidemics took place with an average duration of two years. During the 107 years of the *Black Death* and until 1453, a total of 60 epidemics broke out (although the total number rises to 61 when we add in the Kaffa epidemic).

Based on recorded cases, the *Black Death* appears to have overtaken Europe via the trade routes of the time. The *Second Pandemic* in the Byzantine space presents an interesting distribution since—despite being multi-directional—its directions were not chaotic, but followed a rational course, that is, along the trade routes. The bacterium circulated in the Mediterranean basin, and each outbreak often evolved into a series of epidemics in various parts of the region. Just like in the case of the first pandemic, it should be taken into account that new natural foci emerged, wherefrom new epidemics originated when the risk factors of the disease commingled. In the case of the spread of the *Black Death*, however, it seems that human activity played the most crucial role.

As mentioned before, the spread of the *Black Death* was characterized by *per diem* evolution. The Aegean Sea was a highly busy seaway, where the distances between the coasts and islands are relatively small. The disease must have spread impressively fast during each new outbreak. Observing the aggregations of epidemics, it turns out that the spread was not chaotic, but rather ordered. A typical case were the Venetians territories in Greece. When plague broke out, the bacterium was transmitted along the classical routes of Venetian ships, infecting intermediate ports along its way (Corfu, Cephalonia, Methone, Koroni, Chalkida, Crete, Cyprus).[670]

Another cause for the disease being recycled seems to have been the turbulent history of the region, which was tormented by frequent warfare. Table 8 includes some indicative cases of warfare conducted in the midst of epidemics, which to some extent contributed to the spread of the disease.

Apparently, the natural foci that emerged helped to consolidate the disease in the Eastern Mediterranean basin. It is worth noting that after the demise of the Byzantine Empire and until the late fifteenth century, three long periods of epidemics can be identified. The disease proved to be a major threat in the Balkan territories of the emerging Ottoman Empire that succeeded the Byzantine one. More specifically, during the periods 1455–1470, 1476–1482, and 1489–1500, epidemics broke out

---

670 For maritime networks see Ivanov 2013; Preiser-Kapeller 2013 and 2014.

Table 8. War campaigns and epidemics in Byzantium during the *Black Death*.

Epidemics	Geographic areas	War events in Byzantium
1361/64	Constantinople	Ottoman invasion in Thrace
	Crete	Cretan revolution against Venice
1372/76	Northern Greece	Serbs vs. Ottomans
		Byzantines vs. Serbs
	Western Greece	Albanians vs. Serbs
1378/82	Constantinople	Venice vs. Genoa
	Peloponnese	Navarra's mercenaries vs.
		Principality of Achaia
1386/91	Constantinople	Ottoman siege of Constantinople
	Southern Greece	Florentines vs. Catalans
	Peloponnese	Byzantine Civil War
1417/23	Constantinople	Ottoman siege of Constantinople
	Northern Greece	Ottomans vs. Venetians

every year in both the islands and mainland of Greece.[671] The endemicity of the disease caused major epidemics in 1461 (from Greece up to Bosnia), 1466 (from Central Greece up to Constantinople), and 1477 (from Thrace up to Serbia).

On this matter, let us clarify some issues related to the natural foci of the disease. Human activities around natural foci constitute a significant parameter of disease transmission. In the case of an epidemic outbreak in a natural focus resulting from human activity, climatic parameters should be considered as secondary etiological factors influencing the disease.[672] As mentioned above, the relationship between climate changes and plague is complex. Our data show that the disease can be heat-dependent in particular areas and involve factors like rainfall and dry periods. The finding that the disease can emerge in an area after periods of extended drought needs to be corroborated by more solid data and studies for a precise causal relationship to be established. What is probably most definite is that the relationship of drought with plague is essentially indirect. Intense or extended drought (megadrought) leads to the depletion of water resources, thus gradually weakening animals and making them more susceptible to the disease. Animals are forced to move towards different areas in search of water and food, leading populations (such as rodents) to invade enzootic areas where susceptible individuals increase. When these moving rodents are already infected, they enable the disease to appear in new areas. The movements of wild rodents in nature are generally characterized by a 'normal disposal' in the context of daily search for food and water, or of migra-

---

671 Kostis 1995: 336–339; Varlik 2015: 505.
672 Ben Ari 2011.

tions due to abnormal causes like food scarcity, or natural disasters, for example. As for the range of these movements, wild rodents—unlike 'urban' rodents—are able to cover large distances to secure a new habitat.[673]

Despite indications of climate change and plague being related, studies often reach conflicting conclusions. We also have to clarify that the various rodent species play an important role in this peculiar relationship. For example, the prairie dogs of Wyoming seem to become more susceptible during periods of drought.[674] At the same time, a large study of outbreak reports dated from the past four centuries in pre-industrial Europe found a minor correlation between climate and plague.[675]

Although a firmly documented relationship between plague and periods of drought cannot be established, specific findings concerning the epidemics of the *Black Death* in Byzantium shall be presented. We know that the study of tree rings (dendrochronology) can offer indirect information relating to the climate of past centuries. Dendrochronology examines tree rings and the environmental conditions of the past that influenced their formulation. Dating of tree rings can be used to reconstruct the climate of the past (dendroclimatology).[676] In this context, the fluctuation of drought will be briefly presented here in view of its possible correlation with the disease epidemics that occurred in Byzantium. This presentation shall rely on the model of the *Old World Drought Atlas* (OWDA), which is based on the reconstruction of the self-calibrating *Palmer Drought Severity Index* (PDSI) on a 0.5° latitude/longitude grid centered over Europe, North Africa, and the Middle East from A.D. 0000–2012, and draws data from the National Centers for Environmental Information database of the National Oceanic Atmospheric Administration (NOAA, USA).[677] The OWDA draws additional data from the *International Tree-Ring Data Bank* as well as from contributions by the European dendrochronology and dendroarchaeology community. With the help of the OWDA, we are able to create maps that capture summer wetness and dryness over Europe and the Mediterranean Basin. The basic feature of the OWDA lies in the reconstruction of the *Palmer Drought Severity Index* (PDSI), which employs temperature and precipitation data in order to estimate relative dryness. The PDSI is widely used as an index of drought in agricultural crops as well as a soil moisture monitoring index. The PDSI spans from negative (dry) to positive (wet) values and helps in identifying long-term droughts.[678] According to the classification, −4.0 or higher signifies extreme drought, −3.0 to −3.99 severe drought, −2.0 to −2.99 moderate drought, −1.0 to −1.99 mild drought, up to the high (positive) indices of wet conditions.

---

673 Pollitzer 1954: 260.
674 Eads 2016.
675 Yue 2018.
676 Sheppard 2010.
677 Cook 2015.
678 Palmer 1965: Report No45; Ally 1984: 1100–1109; Tegel 2010: 1957–1959.

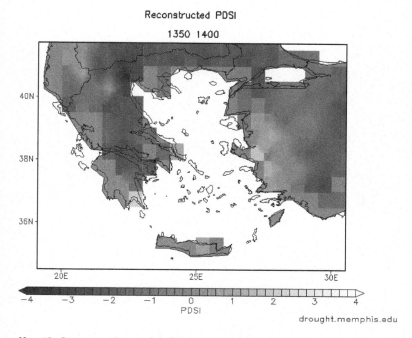

Map 18. Reconstruction model of Palmer Drought Severity Index during the period 1350–1400 (by the author with application of model OWDA[©2015]).

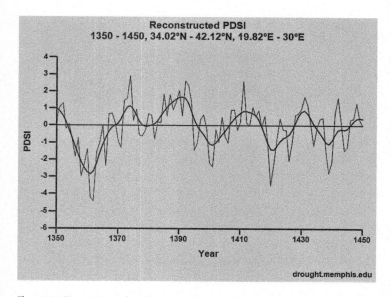

Figure 3. Time-series of PDSI during the period 1350–1400 (by the author with application of model OWDA[©2015]).

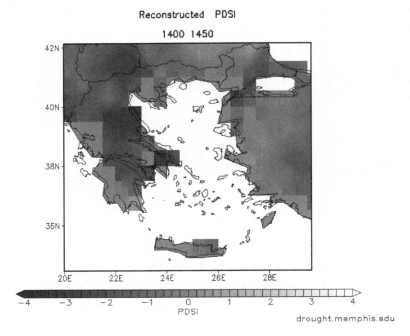

Map 19. Reconstruction model of Palmer Drought Severity Index during the period 1400–1450 (by the author with application of model OWDA[©2015]).

The implementation of this model in Byzantine times reveals some interesting facts. As illustrated in the following map, the area had been affected since the onset of the *Second Pandemic* by long periods of drought, ranging from mild to extreme PDSI (Maps 18 and 19, Figures 3 and 4). Based on the reconstruction of the time, a large percentage of the area including Northern and Central Greece as well as Asia Minor was affected throughout the period 1350–1450. This phenomenon can be characterized as severe drought on the PDSI scale during the period 1400–1450. The distinct periods of drought mainly occurred during the decades of 1360, 1420, and 1430. Despite the lack of more comprehensive data that could contribute to establishing a strong link, it seems that some aggregations of epidemics temporally coincide with the periods of intense negative PDSI values. More specifically, the aggregations of epidemics of 1361–1365 and 1417–1423 appear to concur with extreme PDSI values.

Based on the model employed, severe values were identified in the years 1358 (−3.02) and 1420 (−3.5), whereas the most extreme values were recorded in 1361 (−4.1) and 1362 (−4.4). As mentioned above, it is very difficult to establish a firm correlation. Given that the consequences of drought—which can foster the outbreak of an epizootic that develops into an epidemic—are direct and become evident within a few months, an attempt was made to match the sources with the current year for each PDSI value within the limit or above the range of moderate values. This comparison revealed only two cases of synchronization between epidemics and drought,

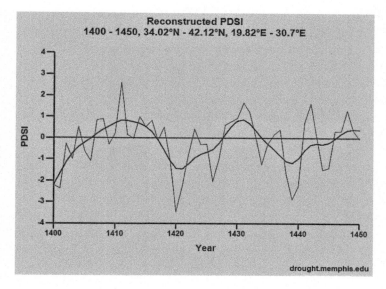

Figure 4. Time-series of PDSI during the period 1400–1450 (by the author with application of model OWDA[©2015]).

which are related to the epidemics and extreme drought of 1361–1362. Although this fact is not sufficient by itself, it should not be overlooked.

The fifteenth century ended with a generalized epidemic on the coasts of Asia Minor and the great *Thanatikon of Rhodes* (1498).[679] At the same time, the Mamluk Sultanate was struck by major epidemics in Egypt (1455, 1459, 1468, 1476, 1492) and Palestine/Syria (1459, 1468, 1476, 1491, 1497).[680] Gradually and up to the nineteenth century, the Westerners regarded the Orient as an extremely dangerous area for contacting infectious diseases, as evidenced by numerous accounts of European travelers and the regulations concerning the quarantine in the western pest-houses (*lazarettos*) imposed to the ships arriving from the ports of the Ottoman Empire.[681]

The Molecular View of the *Second Pandemic*

As is the case in the first pandemic, the place of origin of the *Black Death* has generated a great deal of interest. The rationale behind this interest remains the same, and researchers hope that molecular data might help in solving the mystery. Studies focus on comparing the genomes found in skeletal remains of the period in relation

---

679 Tsiamis 2018b.
680 Kostis 1995: 418.
681 Konstantinidou 2009; Tsiamis 2018a; Chircop 2018.

to the evolution of classification branches. It gradually becomes evident that the micro-evolution of the bacterium plays a crucial role in its overall evolution. For example, in the case of London (1348 – 1350), the findings point towards the 'local' micro-evolution of the invading strain, or even to the presence of multiple strains.[682] The phenomenon of micro-evolution is an acceptable timeless feature, but it poses a problem to modern health services during an epidemic.[683]

The evolution of the microbe shows an increased diversity beyond the boundaries of China.[684] The *Black Death* should be regarded as a 'Big Bang' in the microbe's evolution, since the phylogenetic tree of *Y. pestis* constitutes the basic node of polytomy at the end of branch 0. This view is supported by the analysis of the genomes dating from the era of the *Black Death*, which are closely associated with this node of polytomy.[685] The possible impact of climate changes and the epidemic outbreaks of the *Second Pandemic* should be highlighted here.[686] Perhaps climate changes caused the reduction of susceptible rodents, leading fleas to search for new hosts (humans, caravan camels) traveling along the natural foci of the disease in Asia. It is highly likely that the *Second Pandemic* started from the steppes of Central Asia, possibly during a period of gerbil or marmot epizootic. References to a serious infectious disease can also be traced back to the mid-fourteenth century (1338 – 1339) on the tombstones of the Nestorian graveyards near Issyk Kul Lake in Kyrgyzstan.[687] This is possibly the place where the microbe began its gradual spread to China, India, and Crimea.

Although the majority of researchers consider the Asian steppes to have been the place of origin of the disease, specific discrepancies are found with regard to the biovar. At this point, the issue of the Third Pandemic comes into play. The strains isolated from the areas unaffected by the Third Pandemic belong to the biotype Orientalis.[688] On the other hand, the strains stemming from old enzootic foci belong to the biotypes Antiqua, Medievalis, and Microtus. As far as the biotype Orientalis is concerned, it is thought to have evolved from the biovar Antiqua in China. It is generally hypothesized that the biotype Orientalis is a pandemic type, whereas the biotypes Antiqua, Medievalis and Microtus were local variants with limited epidemic potential.[689] Moreover, the theory that the *Y. pestis* lineage of the *Second Pandemic* emerged independently from rodents into human beings has been proposed.[690] Despite the views put forward in the past, the incredible evolution of molecular tech-

---

682 Bos 2011.
683 Harris 2010.
684 Spyrou 2016.
685 Cui 2016.
686 Schmid 2015.
687 Drancourt 2008.
688 Drancourt 2008.
689 Drancourt 2008.
690 Wagner 2014.

niques has constantly changed the theories on the microbe's evolution. At present, the most prevalent theories have started to question the role of *Y. pestis* Medievalis, and suggest that the biotype Orientalis is responsible for the *Second Pandemic*, or that the *Black Death* originated from distinct clones of *Y. pestis*.[691]

When we examine the phylogenetic evolution, we get the impression that important evolutionary changes occurred in the steppes of China, such as the division of branch 2 into 2.ANT and 2.MED. As for 2.MED, the strains of 2.MED2 and 2.MED3 are found in China, and 2.MED1 in the area of Kurdistan.[692] The polytomy of *Y. pestis* may be further supported not only by molecular findings, but also by the references to epidemic outbreaks in various areas of China, such as the outbreaks that coincided with the Mongol raids. Of course, in a similar fashion to the Byzantine and Arabic sources, the actual nature of the various diseases mentioned in the Chinese sources is not always clear. Epidemics have been recorded in 1213–1222, 1226, 1232, and 1273, whereas, even in the south of the Mongol Empire, epidemics broke out in 1331, 1333, 1344, and 1345.[693]

On the other hand, it appears that the strain that reached Europe began its own evolution, even though the core of its genome remained unaltered. Apparently, this is where the role of the micro-evolution of European strains came into play, as illustrated by the molecular studies of strains from various areas affected by the *Black Death*. A typical example is the study comparing old (already analyzed) strains and strains from skeletal remains found in other archaeological sites. The striking similarities between pair strains i) OSL1 (dating from 1349–1350) from Oslo, Norway and London-ind8124-8291-11972 (1348), and ii) strains SLC1006 (1348) from Saint-Laurent-de-la-Cabrerisse of Southern France and Barcelona 3031 (1348), corroborated historical sources in relation to the spread of the epidemic, while highlighting the vast potential of molecular studies.[694] Besides the impressive findings of molecular studies, however, we should not forget the main principles of classical epidemiology. The role of human activities, movements and wars is verified and well established. Nevertheless, we must bear in mind a basic parameter of the equation, that is urban and wild rodents. At least in the case of Central Europe, it seems possible that Alpine marmots (*Marmota marmota*), a species related to the *Marmota bobak* of Kazakhstan and the *Marmota sibirica* (Manchurian marmot), helped in preserving the microbe.[695] This scenario could be employed and expanded by historians through the study of the proximity of urban centers, as well as of trade and other parameters.

The Eastern Mediterranean comes across as the gateway of a single entry of the disease, or even as the place where the disease was constantly regenerated. After the

---

691 Drancourt 2008; Haensch 2010; Raoult 2013.
692 Morelli 2010.
693 Hymes 2014.
694 Namouchi 2018.
695 Carmichael 2014: 177.

establishment of the disease, indeed, the Mediterranean experienced the reintroduction of strains from infected areas of Western and Central Europe, making it the ideal place for the molecular recycling and micro-evolution of the microbe. Even in the absence of other risk factors, trade routes were by themselves sufficient for the microbe to cross Europe from one end to the other. Unfortunately, the lack of skeletal remains in the Greek region dating from the era of the *Black Death* and subjected to molecular study in view of *Y. pestis*, does not allow us to make comparisons with the respective strains of Western and Central Europe. Although this is a purely hypothetical scenario, it would come as no surprise if such a study concluded that the territories of the Byzantine Empire were affected by multiple strains of *Y. pestis*.

Chapter 9
Theories on the Disappearance of Plague (751–1346)

A major question that has remained unanswered is the reason why plague disappeared from the European continent during the eighth century. The data examined so far show that the dynamics of plague epidemics are influenced by numerous factors. Most theories focus on mutations, fluctuations of immunity in rodent populations, and possible environmental changes.

To approach this issue in a comprehensive manner, we need to consider two significant parameters, which have to be examined together on the basis of the interactive relationship over time: firstly, plague is primarily a zoonotic disease that affects humans when they come into contact with rodents, and secondly, the ability of *Y. pestis* to adapt to its surrounding environment.

With regard to the first parameter, the course and evolution of the *Plague of Justinian* and the *Black Death* have been reconstructed so far, as precisely as possible, with the help of historical sources. The geographic distribution of the epidemics of the first pandemic revealed that the disease was constantly present in the provinces of the Middle East, which can be explained by the existence of enzooty in the area. The broad consensus that plague was absent from Europe between 750 and 1346 is based on indirect assumptions stemming from the absence of historical sources regarding epidemics of such nature and magnitude. However, the inexistence of proof is not necessarily proof of inexistence. Modern experience shows that the needs for information constitute an important criterion for choosing a topic. Small epidemic outbreaks of endemic diseases in Africa rarely appear on the news of the West, and such incidents are usually lost in-between other world events. Of course, national health services nowadays collect epidemiological data made available to international organizations. Accordingly, it is reasonable to assume that the regular outbreak of a disease like plague back in that time gradually led to its integration into people's consciousness, thus reducing it to a common risk. Everybody was aware of the existence of the disease, of the fact that the city where they lived could be struck by plague at any time, but, at the same time, they simply wished that it did not happen. Over the centuries and given the increased appearance of the disease, it should not be considered improbable that the chroniclers of the time stopped being interested in plague. This does not imply, however, that the plague bacterium disappeared from Europe.

When we argue in favor of the possible existence of natural foci in the Byzantine Empire, the question whether an enzootic focus can be deactivated emerges. The classical epidemiological model of plague suggests that the infection of rodents

https://doi.org/10.1515/9783110613636-012

can be transmitted to humans via parasites.⁶⁹⁶ Plague is characterized by long periods of remission in specific natural foci. Environmental changes can lead to alterations in host and reservoir populations, resulting in increased interactions between them and the reemergence of the disease in human populations. This model explains both the onset of sporadic epidemic outbreaks and the persistent presence of natural foci for centuries with the periodic emergence of epizootic in these areas. Various hypotheses have been put forward with regard to the gradual disappearance of the disease, which examine changes in rodents' behavior and changes in the relationship between rodents and humans.⁶⁹⁷ The cause of the gradual disappearance of plague at the urban level is obviously related to the reduction of rodent populations as a result of repeated epizootic. Accordingly, this causes a gradual decline—or even elimination—of infections among the scattered rodents that survive. The decline of rodent populations due to epizootic leads to the natural increase of births during inter-epizootic periods, which results in the preservation of the disease and supports the emergence of the next epizootic. As long as the disease continues to persist among the rodent populations, the latter gradually become highly resistant whereas the fleas do not need to turn to alternative hosts other than rodents. When fleas abandon a dead rodent, they search for a new host among other rodents and not humans. This natural process also facilitates the gradual replacement of susceptible rodents by more resistant ones, the population of which essentially impacts on the frequency of epidemics, from reduction to eventual disappearance.⁶⁹⁸

This is how the disappearance of plague can be explained, at least in the urban centers. Another question arises, though: Could this mechanism account for the deactivation of natural foci, as well? Another parameter comes into play at this point, namely the ability of *Y. pestis* to survive in the free environment, which may be related to the enzootic-epizootic regulatory cycle. The ability of *Y. pestis* to adapt to its surrounding environment is linked to its ability to take advantage of its distinct properties. The most up-to-date and ambitious theory currently available takes as a starting point the moment when the capacities of the established zoonotic epidemiological model are exhausted. This is when the ability of *Y. pestis* to survive in the free environment comes into play, when it is left with no other enzootic options. It is speculated that, when enzooty is impossible, the bacterium can persist in the soil where it develops a bi-directional system of coexistence with the microorganisms and protozoa in the soil, primarily with protozoan cysts.⁶⁹⁹

This approach is based on the phenomenon that occurs in natural foci, which are characterized by long inter-epizootic periods when the bacterium is not found in either hosts or vectors.⁷⁰⁰ This phenomenon cannot be fully explained, and points to

---

696 Drancourt 2006.
697 Appleby 1980.
698 Appleby 1980.
699 Lomaradski 1995.
700 Duplantier 2005.

the hypothesis of a telluric stage of plague, that is, a stage during which the *Y. pestis* bacterium survives in soil thanks to complex relationships that it develops with saprophytes and saprozoa.[701] The top layer of the soil, that is, its surface, bristles with numerous microorganisms. One gram of soil contains millions of bacteria and fungal mycelia, thousands of algae (unicellular plant organisms), and hundreds of protozoa.[702] As long as it remains in the soil, *Y. pestis* is characterized as an autotrophic microorganism that draws energy anaerobically by using inorganic elements like sulfur and iron compounds.[703] The distribution of natural foci is possibly related to the content of the soil in iron and calcium, which are known to be involved in the regulation of the infectious agents of *Y. pestis*, such as the thermo-dependent and Ca^2(+)-dependent gene of the V antigen. In fact, specific plasmids can be found in the genome of the Ca^2(+)-dependent strains of *Y. pestis* as we have seen. Moreover, the iron-acquiring ability of *Y. pestis* is a special characteristic that plays an important role in the bacterium's survival in the host cell. Perhaps its role is equally important in the survival of the bacterium in the natural environment. The role of soil as a natural reservoir of *Y. pestis* has long been under investigation.[704] The telluric forms are able to survive for up to 16 months in soil contaminated *in vitro*.[705] Another source suggests that *Y. pestis* can survive for 1,700 days in the soil, for 400 days in sterile water at 20°C, and for 114 days in non-sterile river water at 18°C.[706]

A recent finding relating to the type of soil that could have possibly allowed *Y. pestis* to survive is particularly interesting. Experimental studies have thus far confirmed that *Y. pestis* survives in soil containing 40 g/L of salt and in hypertonic broth containing up to 150 g/L of salt. When *Y. pestis* is exposed to salt, it creates special round cellular forms (L-forms), a phenomenon known since the era of the microbe's discovery.[707] Moreover, L-forms of *Y. pestis* have been described in natural foci.[708] The microbe has no mechanism to manage exposure to high concentrations of NaCl, but it is able to adapt to gradually increasing salt concentrations. This ability could also be described as the creation of L-forms, in an attempt to gradually adapt to osmotic pressure by decreasing its surface in contact with the environment.[709] A particular finding is that *Y. pestis* survives in saline environments such as the chotts in North Africa, which is the case of all such foci in North Africa as well as other nat-

---

701 Baltazar 1964.
702 Schaechter 2006: 356–370.
703 Schaechter 2006: 356–370.
704 Brygoo and Dodin 1965.
705 Mollaret 1963.
706 Christie 1980; Ayyadurai 2008.
707 Bibel 1976.
708 Zykin 1994; Malek 2017.
709 Malek 2017.

ural foci in Asia with high salt concentrations.[710] Apparently, salt is among the factors helping the survival of *Y. pestis*.

The suspected resistance of *Y. pestis* in soil during inter-epizootic periods has been proposed in the cases of Iran and Madagascar. Although the related mechanism remains unknown, it is suggested that rodents are possibly infected when contacting contaminated soil via inhalation or ingestion, thus triggering a new inter-epizootic cycle, since several studies have shown that *Y. pestis* can survive for at least 24 days in natural conditions.[711] Nevertheless, the virulence of *Y. pestis* gradually decreases over this period.

The telluric bacterium theory combined with the existing epidemiological model and the human ectoparasites—as a factor that determines the onset of the disease— offers the advantage of classifying past and future plague epidemics into six possible scenarios:[712]

1. Epizootic in wild rodents in the absence of human ectoparasites, causing sporadic outbreaks among humans.
2. Epizootic in wild rodents in the presence of human ectoparasites, causing sporadic, limited epidemic outbreaks.
3. Epizootic in wild rodents and mice causing a slow but steady spread of plague in animal populations, accompanied by sporadic outbreaks among humans living in the specific natural focus.
4. Epizootic in wild rodents and mice in the presence of human ectoparasites, causing outbreaks in small groups of people which gradually spread to larger populations.
5. Typical mice plague, as in ports, causing numerous outbreaks in humans with a limited geographic scope and duration.
6. Mice plague in the presence of human ectoparasites, causing extensive plague epidemics.

The possible fading of enzootic natural foci at the end of the *Plague of Justinian* would have logically brought about the disappearance of the bacterium. Apart from hypotheses regarding the gradual disappearance of the disease, however, which involve changes in the relationship between rodents and humans, the hypothesis of cross-immunity against other diseases, such as yersinioses, comes into play. This could possibly offer rodents immunity and weaken enzooty.[713]

Judging from the modern trends of the genus *Yersinia*, we could argue that something similar may have occurred during the eighth century. *Y. enterocolitica* is nowadays considered to be gradually taking over the nosological spectrum of *Y. pseudo-*

710 Malek 2017.
711 Andrianaivoarimanana 2013.
712 Marfat 1998.
713 Fukushima 2001.

tuberculosis.[714] As a result, certain animal species susceptible to plague acquire immunity, whereas the more frequent onset of other—equally infectious—species of the genus *Yersinia* offers a long-lasting immune protection against plague.[715] This hypothesis could help to explain another phenomenon too, namely the cases of subclinical bacteremia in various types of reservoirs, which are usually characterized as resistant to plague while functioning as infectious sources at the same time.[716]

Summing up the phenomenon of the disappearance of plague during the eighth century, it could be argued that the natural foci that emerged during the first two centuries of the first pandemic followed the established course of every natural focus of modern times. The gradual succession of susceptible rodents by more resistant ones resulted in the gradual reduction in the frequency of epizootic outbreaks, and the elongation of inter-epizootic periods and reduced enzooty. If that happened, the *Y. pestis* bacterium temporarily turned to a different means of survival by developing a distinct, telluric form until the conditions became ideal to start a new enzootic cycle.

Finally, the contribution of microorganisms that are symbiotic with *Y. pestis* is a particularly important issue. The finding that the coexistence of *Y. pestis* and *Mycobacterium tuberculosis* does not affect their parallel growth, coupled with the fact that *Mycobacterium lepraemurium* causes leprosy in rodents, but at the same time protects them from plague, raises the issue of identifying the antagonists and symbiotes of *Y. pestis*. To date, it has been possible to define two groups, antagonists and symbiotes, which only include a small number of identified microorganisms. Concerning the antagonism or symbiosis of *Y. pestis* with other microorganisms, however, it has not been clarified whether it is actual antagonism in relation to the metabolic products of bacteria, which cannot be used as nutrient substrates by *Y. pestis*. Therefore, the possibility remains that, during the centuries of its disappearance, *Y. pestis* was restricted by other bacteria until it developed a specific defence mechanism against them, so this phenomenon could serve as a basis for future research.

---

714 Alonso 1999.
715 Malek 2017.
716 Baltazar 1963; Levi 1997.

Epilogue

The present study considered the whole spectrum of plague epidemics that occurred in the Byzantine Empire, from the year of its foundation until its fall, that is, the period between 330 and 1453 A.D. It did not only highlight particular concerns, but it also identified fields for future research and reached a number of conclusions.

The Byzantine Cities as 'Hosts' of Plague

Byzantium went through cycles oscillating between glory and disrepute. The geopolitical shifts during its existence resulted in numerous changes in its demography. At various phases of its history, the Byzantine Empire presented a multiethnical motley. New ethnic groups were constantly incorporated into the Empire through the centuries, and such massive inflow automatically translated into urbanization towards the major centers, while imposing interventions that ultimately led to burdening the health status of cities. The inadequate design of a city, combined with its overpopulation, was bound to catalytically influence the rapid spread of the two pandemics. Furthermore, changes in the dynamic population system through the influx of new peoples, led to fluctuations in the collective immunity of an area as well.

The Difficulty of Identifying the Disease

Based on the study of plague epidemics in the Middle Ages, a number of issues emerged which demonstrated the difficulty of identifying the disease through historical sources. The correct diagnosis of plague in previous centuries must be based on a set of enlightening data with regard to the epidemiological cycle of the disease and its intensity, as well as its periodicity. Differential diagnosis, also, is problematic—since during the early days of disease outbreaks diagnosis is difficult—considering that various infectious diseases are accompanied by similar clinical symptoms and signs.

The descriptions by contemporary chroniclers are often characterized by a sense of pessimism, insecurity, and despair. The perpetuation of this literary model of hyperbole gradually led to the creation of a particular manner of describing the disease. In this type of narration, another problem lies in capturing the view of the time according to which the emergence of plague in a city would overshadow or eliminate any other disease. Furthermore, the descriptions of plague epidemics in Byzantium display one more peculiarity: the information from the sources is the product of historical writing by the chroniclers of the period—and not by physicians. This phenomenon is of particular interest, because the texts of major Byzantine physicians dating from the sixth to the fifteenth century contain no reference to plague.

https://doi.org/10.1515/9783110613636-013

Nevertheless, the clinical picture—when described in detail by the Byzantines—corresponds to known symptoms. Indeed, the differences encountered among the various writers are indicative of the differentiated course of the disease among patients, as well as of its various clinical forms. The references of Procopius and Kantakouzenos regarding the effect of *Y. pestis* on the central nervous system of patients range from coma to delirium, and comply with the known two possible courses of plague. In addition, the testimonies of Procopius and Kantakouzenos concerning the swelling of lymph nodes (buboes) or inflammation of the lungs, in conjunction with the other symptoms, lead us to think of the two forms of the same disease, the bubonic and pneumonic.

'Buboes' and 'Carbons' as Points of Differential Diagnosis

Another problem that contemporary scholars are faced with concerns the terminology that the Byzantines employ to describe symptoms. The Byzantine writers convey the medical legacy of ancient Greeks, whereas the terms 'bubo' and 'carbon' ultimately prove more complex than perceived at first glance. Finding and isolating such terms in early Byzantine historiography or medical treatises, arguably evinces the existence of the disease in Europe before the *Plague of Justinian*. Therefore, we need to clarify the evolution of these terms from Antiquity to the Middle Ages. The term *bubo* appears in the Hippocratic *Sixth Book of Epidemics*, where it is not associated with any infectious disease and especially not with plague. In the late Roman period, Rufus of Ephesus (1st–2nd cent. A.D.), in turn, cites Dioscorides (1st cent. A.D.) and Posidonius (ca. 135-ca. 51 B.C.), who also speak of buboes that were extensively debated during an epidemic in Libya. The fourth-century A.D. physician Oribasius reproduced a text of Rufus entitled *On Buboes* in one of his works. Based on the reference to Rufus, which is currently known thanks to Oribasius, the medical view of the time is that ordinary buboes are located in the neck, armpits and thighs, either accompanied by fever or not, while buboes related to infections are called 'infectious buboes' and they are deadly and virulent. This sort of buboes, however, is necessarily examined in light of the differential diagnosis of bubonic plague and constitutes no *a priori* evidence of the disease. If the so-called 'infectious buboes' in early Byzantine texts are indeed related to the presence of plague, then, the statement by Leo the Philosopher (ca. 790–869 A.D.), who lived after the first pandemic, that plague can be cured just like an inflammation is rather surprising. It is obvious that the term 'bubo' stood for something different in the medical literature of the time.

Another problem in the interpretation of the sources relates to the term 'carbon'. As already mentioned in relation to the *Antonine Plague*, the term is to be traced back to Galen (129-after [?] 216 A.D.) and echoes the general perception of the time regarding the harmful consequences of the imbalance of humours in the human body: 'carbon' results from the accumulation of melancholic humor (black bile) on the skin, or when black and warm blood accumulates, thereby creating a crust. The physicians and chroniclers of the period employ the terms 'bubo' and 'carbon' in order to describe disease entities, and not specific symptoms. The term 'carbon' did not refer to the modern known disease *anthrax*; it described, rather, the broader dermatological manifestation of various diseases—not necessarily infectious—, since even common hardships could give rise to 'carbons'.

The Belief of Plague Existence Before the 6th Century A.D.

The present study also raised the issue of the identification of mysterious epidemics in mythology in both Antiquity and Late Antiquity, which have been improperly considered as either plague or its precursors. The view that the Homeric *Plague of the Achaeans*, the Biblical *Plague of the Philistines*, the *Plague of Athens* (*Plague* or *Syndrome of Thucidides*), the Antonine plague (also known as *Plague of Galen*) and the *Plague of Cyprian* were all cases of plague epidemics, has to be reconsidered, given that no diagnosis of plague can be justified on the basis of the data derived from descriptions. The *Antonine Plague* and the *Plague of Cyprian* were perhaps caused by the same pathogen, but the diagnosis of actual plague is unlikely. The only evidence provided by the above epidemics is a rough recording of the level of collective immunity at the transition from the Ancient to the Medieval world. The cases of epidemics that we have mentioned above further reinforce the view that those who were affected were immunologically virgin populations. This was due to the absence of relevant infectious agents up to that moment, or to the low level of pre-existing collective immunity against certain infectious diseases. Reduced collective immunity may be attributable to the long absence of the infectious agent or to its antigenic drift. After all, it is well known that the level of immunity deteriorates over time, whereas the composition of the population may vary as well, through the introduction of new susceptible individuals who—to a large extent—determine the epidemiological features of a disease regardless of the nature of the infectious agent and the ways of dispersion.

Since the foundation of Constantinople in 330 A.D. and until the appearance of plague in 542 A.D., a series of evidence of epidemics survives; however, the determination of the diseases is difficult due to the lack of description of the symptomatology. Yet, regardless of how nebulous the descriptions might be, they do not direct our hypothetic diagnosis towards plague, but rather to other disease entities, such as smallpox, pharyngeal or laryngeal diphtheria, dysentery, or group poisonings, for example. Non-extensive reports about these epidemics and their symptoms may be attributable to a lack of knowledge by the writers; otherwise, these epidemics were ordinary outbreaks of endemic diseases, which could not be linked to plague.

Recent studies examining numerous strains reveal the presence of the bacterium's sequence in Asia and Europe from the late Stone Age and the Early Bronze Age. It has been argued that the bacterium reached the European continent sometime between 3000–5000 BP. The first estimate, however, is that these strains have no direct relation to the one that affected the Byzantine Empire during the First Pandemic.

The Port of Clysma as a Potential Gateway of Plague in the 6th Century

Byzantine texts define the source of the first pandemic as the wider area of Northeast Africa, in the region of Ethiopia or Egypt. The Egyptian city of Pelusium on the Mediterranean coast went down in history as the city where the epidemic began in 541 A.D. Although the outbreak of the disease in that particular city is undeniable, questions arise as to the actual gateway of the disease into Byzantine Egypt. The case of plague arriving in Pelusium through an endemic or epidemic source in the Mediterranean cannot be taken for granted, due to the absence of references to a similar disease before 541. We thus must turn to the possible overland arrival of plague in Pelusium. With the help of historical sources, the land and maritime trade routes of the Egyptian inland, along with the maritime commercial network of the Red Sea, were reconstituted here. A series of historical events, such as the involvement of the Persians and Byzantines in a local conflict between the kingdoms of Aksum (Ethiopia) and the Homerites (Yemen), caused intense political, military, and commercial disputes in the area of the Red Sea. The Aksumites protected Byzantine trade in the area, while, at the same time, they took over the retail trade of the Byzantine Empire with India. Our study suggests Clysma, a port on the Red Sea at the entrance of modern Suez Canal on the Sinai Peninsula, as a potential gateway of the disease. Based on the network of canals of the Nile, we speculated that the disease was transferred from Clysma to Pelusium. The historical role and commercial importance of Clysma, the specific time of outbreak of the epidemic in Pelusium—which coincides with the rise of the water level of the Nile—, and the operation of Trajan's Canal—which linked Clysma with the Nile—, are all evidence that supports this hypothesis. The general view that the Trajan Canal was out of order during the sixth century is being questioned. A careful reading of the report by Gregory of Tours on the existence and operation of a canal in Clysma at that period would be enough to alter the facts that we know of today, whereas it surely strengthens the case that this port functioned as a gateway of the disease into the province of Egypt.

The Problem of the Origin of the *First Pandemic*

The proposal of Clysma as a potential gateway of the disease opens a different perspective to the question of the place of origin of plague. The two prevailing views refer to Central Africa and Asia respectively. The studies of the strain that caused the first pandemic indicate a relationship between the genomes of branch 0.ANT and the strains found in the Xingjiang region in north-western China, or in Kyrgyzstan. All strains of the first pandemic appear to originate from the second/third-century A.D. genome found in the region of the Tian Shan mountain range in Central Asia, whereas the first indications suggest it is the place of origin of branch 0.ANT as the ancestor of high-virulence strains. The theory of African origin, on the other hand, seems less likely nowadays. Nevertheless, studies point to a close relationship between the strain Angola and the Pestoides group, thus making it one of the ancestral strains within the *Y. pestis* lineage.

Redefining the Epidemics of the *First Pandemic*

The present work attempts to redefine existing studies on the epidemics that struck the Byzantine Empire in the period 541–750 A.D. Prior to the grouping of epidemics, we explained the problematic meaning of the term 'wave' and the reason why it needed to be redefined. By examining the evolution of the pandemic from a chronological and geographical perspective, we were able to identify 15 aggregations of epidemics, and 16 local outbreaks in the frame of the first pandemic in Byzantium.

According to existing grouping methods, it seems that plague appeared in the major urban centers of the Empire every 10–15 years. The recording of epidemics between the sixth and seventh century showed a geographical orientation of the disease towards the provinces of Syria, Mesopotamia, Italy, and Constantinople. By the late seventh century and during the eighth, an upward movement of the disease had became apparent, not only in Syria, but also in Mesopotamia and Egypt, while the opposite was true for Italy and Constantinople, where a gradually downward trend was noticed. During the 208 years of the first pandemic, 79 epidemics took place with an average duration of two years.

Data Supporting the Possible Existence of a Natural Focus in Syria

The grouping methods of the epidemic aggregations of the first pandemic that were applied in the past focused on the time of the outbreak and its geographical distribution. Nevertheless, human activities such as trade and movements of troops remained the centerpiece of epidemics. The present study evaluated the periodic appearance of the disease in light of the possible existence of enzootic regions within the Empire, without rejecting, however, the significance of the aforementioned factors. The theory of development of enzootic foci in some provinces of the Empire, which occasionally caused epidemics under the influence of climatic and environmental conditions, appears to be realistic.

After grouping these epidemics, we turned our attention to a specific region of the Empire, the Middle East. As we said, Palestine, Mesopotamia, and particularly Syria were the areas where most epidemics broke out during the *Plague of Justinian*. We oberved a permanent presence of plague in the province of Syria in these regions, whereas the disease seems to have affected the border of North Syria with Mesopotamia, in the area of the 'Dead Cities' in the period 542–610 A.D.

The behavior of plague in the wider area is reminiscent of the dynamics of modern outbreaks of epidemics. Outbreaks occurred in areas where the disease had long been absent, and forgotten foci were re-activated after decades, with the epidemics of Antioch being the most typical example. It seems that the area between Lake Tiberias, the Golan Heights, and Damascus experienced the most outbreaks of epidemics. It seems also that plague was recycled among Syria, Mesopotamia, and Palestine, which constituted a source of constant danger. It can be argued that, after the first aggregation of 541–546, some areas of Syria had become permanently enzootic, periodically becoming epizootic and experiencing outbreaks. Phenomena like the outbreak of an epidemic in a city or an area where it had been absent for decades, may be explained by the sudden invasion of infected animals, as well as by the sudden activation of natural foci that had been inactive for decades, in a similar way to modern epidemics. Unfortunately, the epizootic role of rodents is often bypassed. The enzootic behavior of rodent populations allows the disease to be maintained for years; the longer this relationship between disease-resistant and non-resistant species, the longer plague remains active.

Another element that reinforces the view of enzootic areas in Byzantine provinces is the emergence of epidemics after natural disasters. The study of historical sources indicated five cases of epidemics in Byzantium, in a period ranging from a few months to a whole year after an earthquake. Accordingly, the population of rodents gradually rises over the next eight to ten months after an earthquake, at which point the balance between them and their ectoparasites gets disrupted.

Identifying the Origin of the *Black Death*

The sudden re-emergence of the *Black Death* has been the subject of extensive medical and historical research. The narrative by de' Mussi is a valuable source, but it is of little help from an epidemiological perspective. Up to the present, we have been aware of the obvious only, namely that the disease spread from Kaffa to Constantinople. This study revealed an alternative course, with Trebizond being the possible first victim of the *Second Pandemic*. Our interest now goes beyond the purely historical perspective and any attempts to provide answers take into account the findings of molecular studies. Our previous knowledge in the field of molecular research hinted at the association of the three strains of *Y. pestis* (Antiqua, Medievalis, and Orientalis) with the respective pandemics (*Plague of Justinian*, *Black Death*, and *Third Pandemic*). The most recent molecular data seem to repudiate some of the previous data, and suggest that the pandemic biotype Orientalis is responsible for the *Black Death*. In any case, it is certain that the *Black Death* favored the microbe's evolution from a phylogenetic perspective.

The Arabic chroniclers of the fourteenth century mention epidemic outbreaks in the wider region of the Caspian Sea. The Mongol conquest and the unification of the tribes of Eurasia under Genghis Khan in the thirteenth century resulted in the restoration of commercial relations with Southeast Europe and the Arabic world. This transformation is considered instrumental in the movement of *Y. pestis* to the West and the gradual passing of plague to nosologically virgin populations. Most scientific approaches embrace the theory that plague originated in the high plateaus of Central Asia. A few years prior to the spread of the microbe in the Black Sea region, the area of Issyk Kul in Kyrgyzstan seemingly played a catalytic role in the spread of the disease.

An alternative course has been proposed in this study, taking into account the epidemiological characteristics of enzootic wild rodents of West Asia, as well as historical data. To further the view that supported a possible existence of natural enzootic foci in the regions of Syria and Mesopotamia, we should consider a gradual transfer of *Y. pestis* from the Middle East to the East as probable. Presumably, this transfer began gradually, after the conquest of Damascus by the Mongols in 1258 and their consequent contact with the natural foci mentioned above. According to the alternative course suggested here, plague must have gradually spread north to the Caspian Sea and the Caucasus. Then, it moved east, through North Iraq and Kurdistan to the steppes of Central Asia. A crucially significant role in this movement was played by the enzootic rodents of the Caucasus, which infected other species of rodents in West Asia—such as marmots and gerbils—, thus making these areas enzootic.

This scenario is merely hypothetical, although it might prove more valuable in light of potential molecular findings from the strains of the Middle East. Maybe then we could have a clearer picture of the potential relationship of these strains with those of the Caucasus and the steppes of Central Asia, all the more so given the references of unknown epidemics in the Middle East several centuries before the reemergence of plague in Kaffa. The area of the Middle East is considered as the 'black box' of plague history and the missing link in the evolution of the microbe. The potential future discovery of skeletal remains with molecular indications of infections with *Y. pestis* would definitely foster our understanding of the molecular history and geography of the disease.

Apparently, *Y. pestis* managed to survive in the difficult environmental conditions of the area by taking advantage of a special physiological mechanism, that is, the hibernation of receptor rodents. In this perspective, we should not exclude the possibility of *Y. pestis* preservation through susceptible populations of rodents, which made the disease enzootic in areas of the Caucasus, long before the first official record of the resurgence of plague in South Russia in 1346.

Clinical Form of the *Black Death*

The clinical form of the *Second Pandemic* has also been an issue of much controversy. Besides any literary resemblance to Thucydides' narrative of the *Plague of Athens*, Kantakouzenos offers a description of the clinical form of pneumonic plague. In line with the prevailing view of the time regarding the contagiousness of the disease through the *miasma*, Kantakouzenos ultimately concluded that the transmissibility of the two clinical forms was possible through contact between people. In Constantinople, however, it soon became evident that no direct contact with the patient was necessary for someone to get infected and sick. The information about the disease being transmitted from person to person through breathing echoes the medieval conception of impure air, while, at the same time, it directs contemporary thinking towards the clinical form of pneumonic plague. However, we should not rush to any definitive conclusion, since various other Byzantine sources of later years speak of the bubonic form. Kantakouzenos' account of the clinical picture involving inflammation, chest pain, and haemoptysis, combined with, and enhanced by, the high mortality and spread rate of the disease, has led many researchers to the diagnosis of its pneumonic form. As far as the mortality rate of the *Black Death* is concerned, bubonic plague ranged between 40 – 60 % of the population, whereas the pneumonic form reached as high as 95 %. These increased rates are related to the fact that plague in the Middle Ages was predominantly 'urban-oriented', in contrast to modern outbreaks of the disease that affect remote rural areas.

Pneumonic plague, however, is considered an unusual form of plague, in that it is the only form transmitted directly from person to person. The present study assumes that bubonic plague may have originally broken out in Constantinople, where individuals with secondary pneumonic plague became infectious and transmitted the disease to their relatives; in turn, these relatives could be held responsible for the direct transmission of plague and the emergence of its primary pneumonic form.

Redefining the Epidemics of the *Black Death*

As with the first pandemic, it was deemed necessary in the case of the *Black Death* to regroup the epidemics of the period. During the *Black Death*, nine aggregations of epidemics were distinguished, along with 16 outbreaks. The geographical distribution of the epidemics indicates that Constantinople, together with its surroundings and the Peloponnese, was the most affected area, followed by Crete, Cyprus, and the islands of the Ionian Sea.

It seems that Constantinople was the city to have been most often affected by the disease, at a frequency of 11.1 years on average. Thessaloniki and Trebizond follow in the preference list of plague with three epidemics, and almost every Venetian acquisition in the Aegean was affected at least once. The maritime network of the Aegean contributed to the spread of the disease to the islands. Venetian Crete, Cyprus and Corfu were harshly affected by plague due to their commercial and strategic position, and the bases of Methone and Koroni were simultaneously affected in every epidemic.

Plague Disappears (751–1346 A.D.)

The new trend among researchers is based on two elements; i) the fact that plague is eminently a zoonosis and epidemics are caused by the human interaction with rodents, and ii) the ability of the bacterium to survive in the free environment. The existing epidemiological model of spread of the infection from rodents to humans through ectoparasites, offers an explanation for both the occurrence of sporadic outbreaks of epidemics and the persistent presence of natural foci for centuries following the enzootic disease in these areas.

Urban-type plague epidemics gradually decreased, due to the reduction of the population of rodents after repeated epidemics. Although reproduction helps stabilize their population, at the same time it leads to the gradual replacement of susceptible species by more resistant ones. This would not only affect the epizootic in a region, but also enzooty. In recent years, a new life-cycle of the bacterium has been proposed, which is activated in inter-epizootic periods or in periods of epizootic absence, and involves the symbiosis with soil organisms and protozoa, thus allowing the characterization of *Y. pestis* as a 'telluric' bacterium. Therefore, the possible depletion of natural enzootic foci at the end of the *Plague of Justinian*, as a result of the gradual replacement of susceptible rodents by more resistant ones, perhaps turned *Y. pestis* into a telluric bacterium. Finally, the issue of the effect of microorganisms antagonistic to *Y. pestis*, which could potentially limit the activity of the bacterium for a certain period of time, remains open to study.

Lists of Tables, Figures and Maps

Tables

Figures

Maps

https://doi.org/10.1515/9783110613636-014

Bibliography

Sources

In the footnotes and the bibliography, some works published without a specific name of author and referred to in this study are identified by either the name of the institution that compiled and edited them (e. g. Centers for Disease Control and Prevention) or the title of the section of the journal where they appeared (e. g. Obituary or Official Publications Received).

Acta Koutloumousiou: P. Lemerle (ed.), *Actes de Kutlumus*. Edition diplomatique (Archives de l'Athos 2). Paris: Lethielleux, 1946 (2nd edition: 1988).

Agapius, *Historia universalis*: A. Vasiliev (ed. and transl.), *Kital al-'unva (Histoire universelle, écrite par Agapius (MahBoub) de Menbidj)*. Editée et traduite en français, Seconde partie II, in R. Graffin and F. Nau (eds), *Patrologia orientalis*, tomus octavus. Paris: Firmin-Didot, 1912, pp. 397–550.

Agathias Myrinaeus, *Historiae*: B. G. Niebuhr (ed.), *Agathiae Myrinaei Historiarum libri quinque* (Corpus scriptorum historiae Byzantinae. Editio emendatior et copiosor, Pars III). Bonn: Weber,1828.

Ammianus Marcellinus: V. Gartdthausen (ed.), *Ammiani Marcellini Rerum Gestarum libri qui supersunt*. Recensuit notisque selectis instruxit, 2 volumes. Leipzig: Teubner, 1874–1875.

Beda, *Historia ecclesiastica*: J. Stevenson (ed.), *Venerabilis Bedae Historia Ecclesiastica gentis Anglorum*. Ad fidem codicum manuscriptorum recensuit. Londini: Sumptibus Societatis, 1838.

Chronica Breviora: Σπ. Λάμπρου (ed.), *Βραχέα Χρονικά*, ed. Κ. Ι. Ἀμάντου (Ἀκαδημία Ἀθηνῶν, Μνημεῖα Ἑλληνικῆς Ἱστορίας, τόμος Αʹ, τεῦχος 1). Ἀθῆναι: Ἀκαδημία Ἀθηνῶν, 1932.

Chronica Byzantina Breviora: P. Schreiner (ed.), *Die byzantinischen Kleinchroniken/Chronica Byzantina Breviora*, 3 volumes (Corpus Fontium Historiae Byzantinae 12, 1–3 Series Vindobonensis). Wien: Österreichische Akademie der Wissenschaften, 1975–1979.

Chronica Ioanninensis: Λ. Βρανούσης (ed.), "Τὸ χρονικόν τῶν Ιωαννίνων κατ' ἀνέκδοτον δημώδη ἐπιτομήν", *Ἐπετηρὶς τοῦ Μεσαιωνικοῦ Ἀρχείου* 12 (1962), pp. 57–115.

Chronica Minora, II, *Chronicon ad annum Domini 846*: E.-W. Brooks (ed), *Chronica Minora*, Pars secunda (Corpus Scriptorum Christianorum Orientalium, Scriptores Syri, Textus, Series tertia, Tomus 4). Paris: E Typographeo Reipublicae, 1904, pp. 157–138; Lat. transl.: I. Guidi, *Chronica minora*, Pars prior (Corpus Scriptorum Christianorum Orientalium, Scriptores Syri, Versio, Series tertia, Tomus 4). Paris: E Typographeo Reipublicae, 1903, pp. 121–180.

Chronica Zuqnīni: A. Harak (ed.), *Chronicle of Zuqnīn*. Pars III and IV: A.D. 488–775 (Mediaeval Sources in Translation). Toronto: Pontifical Institute of Medieval Studies, 1999.

Chronicon Breve: Imm. Bekker (ed.), Chronicon Breve, in Imm. Bekker (ed.), *Ducae Michaelis Ducae Nepotis Historia Byzantina*. Recognovit et interprete Italo addito supplevit (Corpus Scriptorum Historiae Byzantinae. Editio emendatior et copiosior). Bonn: Weber, 1834, pp. 515–525.

Constantinus Harmenopulus: G. E. Heimbach (ed.), *Constantini Harmenopuli, Manuale legum sive Hexabiblos cum appendicibus et legibus agrariis*. Ad fidem antiquorum librorum MSS. editionum recensuit, scholiis nondum editis locupletavit, latinam Reitzii translationem correxit, notis criticis, locis parallelis, glossario illustravit. Leipzig: Weigel, 1851.

Constantinus Porphyrogenitus: G. Moravcsik (ed.), and R. J. H. Jenkins (transl.), *Constantine Porphyrogenitus, De Administrando Imperio*. New, Revised Edition (Corpus Fontium Historiae Byzantinae 1). Washington, D.C.: Dumbarton Oaks Center for Byzantine Studies, 1967.

https://doi.org/10.1515/9783110613636-015

Corpus Hippocraticum: É. Littré, *Œuvres complètes d'Hippocrate*, 10 vols. Paris: J.-B. Baillière, 1839–1861.

Demetrius Cydones, *Epistulae*: J.-F. Boissonade (ed.), Δημητρίου τοῦ Κυδώνη Ἐπιστολαί, in J.-F. Boissonade (ed.), *Anecdota nova*. Descripsit et annotavit. Paris: Dumont 1844, pp. 251–327.

Digesta Iustiniani Augusti: Th. Mommsen (ed.), *Digesta Iustiniani Augusti*, recognovit adsumpto in operis societatum Paulo Kruegero, volume 2. Berlin: Weidmann, 1870.

Egeria, *Itinerarium*: P. Maraval (ed.), *Egérie, Journal de voyage (Itinéraire)*. Introduction, texte critique, traduction, notes, index et cartes (Sources chrétiennes 296). Paris: Cerf, 1982.

Elias Nisibenus, *Opus Chronologicum*: E. W. Brooks (ed.), *Eliae Metropolitae Nisibeni Opus Chronologicum*. Pars prior (Corpus Scriptorum Christianorum Orientalium, Scriptores Syri, Versio, Series tertia, Tomus VII). Romae: Karolus De Luigi; Parisiis: Carolus Poussielgue; Lipsiae: Otto Harrassowitz, 1910.

Eusebius, *Historia Ecclesiastica*: J.-P. Migne (ed.), *Eusebii Pamphili Ecclesiastica Historia*, in J.-P. Migne (ed.), *Eusebii Pamphili, Caesareae Palaestinae Episcopi, Opera omnia quae exstant, curis variorum ...*, Tomus secundus (Patrologiae Graecae cursus completus ..., Series Graeca, Patrologiae graecae tomus XX). Paris: J.-P. Migne, 1857, cols. 45/46–906/906.

Eusebius, *Hieronymi Chronicon*: R. Helm (ed.) *Eusebius Werke*, Band 7, Erster Teil: *Die Chronik des Hieronymus* (Die griechischen christlichen Schriftseller der Ersten Drei Jahrhunderte 27). Leipzig: J. C. Hinrichs, 1913.

Evagrius, *Historia Ecclesiastica*: J. Bidez, and L. Parmentier (eds), *The Ecclesiastical History of Evagrius with the Scholia*. Edited with Introduction, Critical Notes, and Indices. London: Methuen, 1898.

Galenus: K. G. Kühn (ed.), *Claudii Galeni Opera omnia* (Medicorum Graecorum Opera omnia quae exstant 1–20), 22 volumes. Leipzig: Knobloch, 1821–1833.

Galenus, *De differentiis febrium*: K. G. Kühn (ed.), *Galenus, De differentiis febrium*, in K. G. Kühn (ed.), *Claudii Galeni Opera omnia*, vol. 7 (Medicorum Graecorum Opera omnia quae exstant 7). Leipzig: Knobloch, 1824, pp. 273–405.

[Galenus], *Definitiones medicae*: K. G. Kühn (ed.), *Galenus, Definitiones medicae*, in K. G. Kühn (ed.), *Claudii Galeni Opera omnia*, vol. 19 (Medicorum Graecorum Opera omnia quae exstant 19). Leipzig: Knobloch, 1830, pp. 346–462.

Galenus, *De libris propriis*: I. Mueller (ed.), *Claudii Galeni Pergameni De libris propriis* (Claudii Galeni Pergameni Scripta minora II). Leipzig: B. G. Teubner, 1891, pp. 91–124.

Georgius Sphrantzes, *Chronica*: R. Maisano (ed. and transl.), *Georgio Sfranze, Cronaca*. Saggio introduttivo, edizione, traduzione e note (Corpus Fontium Historiae Byzantinae, Series Italica 29; Scrittori bizantini, 2). Rome: Accademia Nazionale dei Lincei, 1990.

Gregorius Nyssenus: J.-P. Migne (ed.), *S.P.N. Gregorii Episcopi Nysseni Tractatus Secundus in Psalmorum Inscriptiones*, in J.-P. Migne (ed.), *S.P.N. Gregorii Episcopi Nysseni Opera quae reperiri potuerunt omnia (Ed. Morell. 1638) nunc denuo correctius et accuratius edita et multis aucta*, Tomus primus complectens scripta exegetica (Patrologiae Graecae cursus completus ..., Series Graeca Prior, Patrologiae graecae tomus XLIV). Paris: J.-P. Migne, 1863, cols 487/488–615/616.

Gregorius Papa, Registrum Epistularum: D. Norberg (ed.), *Gregorii I Papae Registrum Epistularum Libri I-VII*, 2 volumes (Corpus Christianorum, Series Latina 140–140 A). Turnhout: Brepols, 1982.

Gregorius Turonensis, Historiarum Libri: B. Krusch, and W. Levison (eds), *Gregorii Episcopi Turonensis Opera*, 1. *Libri Historiarum X*. Editio altera (Monumenta Germaniae Historica, Scriptores Rerum Merovingicarum, Tomus 1, Pars 1, Fasciculi I-III). Hannover: Hahen, 1937.

Gregorius Turonensis, Miracula: B. Krusch (ed.), *Georgii Florentini Gregorii Episcopi Turonensis Libri octo Miraculorum*, in B. Krusch (ed.), *Gregorii episcopi Turonensis Miracula et Opera minora* (Monumenta Germaniae Historica, Scriptores Rerum Merovingicarum, Tomus 1, Pars 2). Hannover: Hahen, 1885, pp. 1–294 (reprint 1969).

Historia Augusta: D. Magie (ed. and transl.), *Scriptores Historiae Augustae*, 3 volumes (Loeb Classical Library 139, 140, 263). Cambridge, MA, and London: Harvard University Press, 1921, 1924, 1932.

Ibn Battuta, Rihla: C. Defrémery, and B. R. Sanguinetti (eds and transl.), *Voyages d'Ibn Batoutah*. Texte Arabe, accompagné d'une traduction, 4 volumes (Collection d'ouvrages orientaux publiée par la Société Asiatique). Paris: Imprimerie Impériale, 1853–1858.

Joannes Caminiates, *De expugnatione Thessalonicae*: G. Böhlig (ed.), *Ioannis Caminiatae De expugnatione Thessalonicae* (Corpus Fontium Historiae Byzantinae, Series Berolinensis 4). Berlin, and New York, NY: De Gruyter, 1973.

Joannes Cantacuzenus, *Historiae*: L. Schopen (ed.), *Ioannis Cantacuzeni ex imperatoris Historiarum Libri IV.* Graece et Latine. Volumen III (Corpus Scriptorum Historiae Byzantinae. Editio emendatior et copiosior). Bonn: Weber, 1832.

Joannes Chrysostomus: J.-P. Migne (ed.), *S. P. N. Joannis Chrysostomi Homiliae XC in Matthaeum* (*S. P. N. Joannis Chrysostomi, Archiepiscopi Constantinopolitani, Opera Omnia quae exstant, vel quae eius nomine circumferuntur*, ad MSS. Codices gallicos vaticanos, anglicos, germanicosque; necnon ad Savilianam et Frontonianam editiones, castigata, innumeris aucta; nova interpetatione ubi opus erat, praefationibus, monitis, notis, variis lectionibus illustrata; nova sancti doctoris vita, appendicibus, onomastico et copiosissimis indicibus locupletata opera et studio D. Bern. De Montfaucon, monachi Benedictini e congr. S. Mauri, editio novissima, iis omnibus illustrata quae recentius tum Romae, tum Oxonii, tum alibi a diversis in lucum primum edita sunt, vel jam edita, ad manuscriptorum diligentiorem crisim revocata sunt. Tomi septimi, pars prior et Tomi septimi, pars posterior (Patrologiae Graecae cursus completus, Series Graeca Prior, Patrologiae graecae tomus LVIII). Paris: J.-P. Migne, 1862.

Joannes Malalas, *Chronographia*: L. Dindorf (ed.), *Ioannis Malalae Chronographia*. Accedunt Chilmeadi Hodiique annotationes et Ric. Bentleii Epistola ad Io. Millium (Corpus Scriptorum Historiae Byzantinae. Editio emendatior et copiosior). Bonn: Weber, 1831.

Joannes Tzetzes, *Epistulae*: P. Leone (ed.), *Ioannis Tzetzae Epistulae* (Bibliotheca scriptorum Graecorum et Romanorum Teubneriana). Leipzig: Teubner, 1972.

Laonicus Chalcocondyla, *Historiae*: Imm. Bekker (ed.), *Laonici Chalcocondylae Atheniensis Historiarum Libri decem* (Corpus Scriptorum Historiae Byzantinae. Editio emendatior et copiosior). Bonn: Weber, 1843.

Leontius, *Vita Sancti Joannis*: A.-J. Festugière (ed.), *Leontios de Néapolis, Vie de Syméon le Fou et Vie de Jean de Chypre*. Edition commentée en collaboration avec L. Rydén (Institut français d'archéologie de Beyrouth, Bibliothèque archéologique et historique 95). Paris: Librairie orientaliste Paul Geuthner, 1974.

Leontius Machairas, *Chronaca*: K. N. Σάθας (ed.), *ΕΞΗΓΗΣΙΣ τῆς γλυκείας χώρας Κύπρου, ἡ ὁποία λέγεται Κρόνακα τουτέστιν Χρονικόν* (Μεσαιωνικὴ Βιβλιοθήκη 2). Βενετία: Τύποις τοῦ Χρόνου, 1873.

Liber Pontificalis: L. Duchesne (ed.), *Le Liber Pontificalis*. Texte, Introduction, Commentaire. Tome Premier (Bibliothèque des Ecoles françaises d'Athènes et de Rome, 2ᵉ série, 2). Paris: Ernest Thorin, 1886.

Marcellinus Comes, *Chronicon*: Th. Mommsen (ed.), *Marcellini V. C. Comitis Chronicon ad A. DXVIII Continuatum ad A. DXXXIV cum additamento ad A. DXLVIII*, in Th. Mommsen (ed.), *Chronica minora saec. IV, V, VI, VII* (Monumenta Germaniae Historica, Auctores Antiquissimi 11). Berlin: Weidmann, 1894, pp. 37–108.

Michael Syrianus, *Chronica*: J.-B. Chabot (ed.), *Chronique de Michel le Syrien, Patriarche Jacobite d'Antioche (1166–1199)*. Editée pour la première fois et traduite en français, 4 vols. Paris: E. Leroux, 1899–1910.

Miracula Sancti Artemii: Ath. Papadopoulos-Kerameus (ed.), "Διηγήσις τῶν θαυμάτων τοῦ ἁγίου καὶ ἐνδόξου μεγαλομάρτυρος καὶ θαυματουργοῦ Ἀρτεμίου", in Ath. Papadopoulos-Kerameus (ed.), *Varia Graeca Sacra: Sbornik grečeskich neizdannych bogoslovskich tekstov IV-XV vekov* (Zapiski Istoriko-Filologičeskago Fakulteta Imperatorskago S.-Peterburgskago Universiteta Istoriko-Filologičeskij Fakultet 95). Sankt Petersburg: V. F. Kirschbaum, 1909, pp. 1–75.

Nicephorus Constantinopoli Patriarchus, *Breviarium*: C. Mango (ed. and transl.), *Nikephoros, Patriarch of Constantinople: Short History* (Dumbarton Oaks Texts 10; Corpus Fontium Historiae Byzantinae 13). Washington, D.C.: Dumbarton Oaks, 1990.

Nicephorus Gregoras, *Byzantina Historia*: L. Schopen (ed.), *Nicephori Gregorae Byzantina Historia Graece et Latine cum annotationibus Hier. Wolfii, Car. Ducangii, Io. Boivini et Cl. Capperonnerii*, 2 volumes (Corpus Scriptorum Historiae Byzantinae. Editio emendatior et copiosior). Bonn: Weber, 1829–1830.

Oribasius, *Collectiones medicae*: J. Raeder (ed.), *Oribasii Collectionum Medicarum Reliquiae*, volume 3: *Libri XXIV-XXV.XLIII-XLVIII* (Corpus Medicorum Graecorum VI 2,1). Lipsiae et Berolini: B. G. Teubner, 1931.

Paulus Diaconus, *Historia Langobardorum*: L. Bethmann, and G. Waitz (eds), *Pauli historia Langobardorum*, in G. Waitz (ed.), *Scriptores rerum Langobardicarum et Italicarum saec. VI-IX.* (Monumenta Germaniae Historica inde ab anno Christi quingentesimo usque ad annum Germanicarum Medii Aevi). Hannover: Hahn, 1878, pp. 12–187.

Philostorgius, *Historia Ecclesiastica*: J. Bidez (ed.), *Philostorgius Kirchengeschichte mit dem Leben des Lucian von Antiochien und den Fragmenten eines Arianischen Historiographen* (Die griechischen Schriftsteller der ersten drei Jahrhundert 21). Leipzig: J. C. Hinrichs, 1913.

Procopius, *De aedeficiis*: J. Haury (ed.), *Procopii Caesariensis. Opera Omnia*, volume 3.2: *VI Libri ΠΕΡΙ ΚΤΙΣΑΜΑΤΩΝ sive De aedificiis cum duobus indicibus et appendice* (Bibliotheca Scriptorum Graecorum et Romanorum Teubneriana). Leipzig: B.G. Teubner, 1913.

Procopius, *De bellis*: J. Haury (ed.), *Procopii Caesariensis. Opera Omnia*, volume 1: *De Bellis. Libri I-IV* (Bibliotheca Scriptorum Graecorum et Romanorum Teubneriana). Leipzig: B. G. Teubner, 1905.

Pseudo-Dionysius, *Chronicon*: J.-B. Chabot (ed.), *Incerti Auctoris Chronicon Pseudo Dionysianum Vulgo Dictum*, vol. II. *Accedunt Johannis Ephesini fragmenta* (Corpus Scriptorum Christianorum Orientalium 104 = Scriptores Syri 3.2, Textus). Parisiis: e Typographeo Reipublicae, 1933.

English translation: W. Witakowski, *Pseudo-Dionysios of Tel-Mahre, Chronicle (known also as the Chronicle of Zuqnin), Part III.* Translated with notes and introduction (Translated Texts for Historians 22). Liverpool: Liverpool University Press, 1996.

Severus: B. Evetts (ed. and transl.), *Severus of Al' Ashmunein (Hermopolis), History of the Patriarchs of the Coptic Church of Alexandria*, Volume 3: Agathon—Michael I (766). Arabic text edited, annotated, and translated (Patrologia Orientalis 21 [5.1]). Paris: Firmon-Didot, 1907.

Theodosius Melitenus, *Chronographia*: Th. L. Fr. Tafel (ed.), *Theodosii Meliteni qui fertur Chronographia ex Codice Graeco Regiae Bibliothecae Monacensis*, edidit et reformavit (Monumenta saecularia Academiae Scientiarum Monacensis III. Classe, 1. Heft). Munich: G. Franz, 1859.

Theophanes, *Chronographia*: J. Classen (ed.), *Theophanis Chronographia*, volume 1 (Corpus Scriptorum Historiae Byzantinae. Editio emendatior et copiosor) Bonn: Weber, 1839.

Theophylactus Simocatta, *Historiae*: Imm. Bekker (ed.), *Theophylacti Simocattae Historiarum. Libri octo* (Corpus scriptorum historiae Byzantinae. Editio emedatior et copiosor). Bonn: Weber, 1834.

Thucydides, *Historiae*: C. Hude (ed.), *Thucydides Historiae*. Volume I: *Libri I-IV.* Lipsiae: Teubneri, 1901.

Tura del Grasso, *Chronicon Senense*: L. Muratori (ed.), *Chronicon Senense, Italice scriptum ab Andrea Dei, et ab Angelo Turae continuatum, exordium habens ab Anno MCLXXXVI. & desidens in Annum MCCCLII. è Manuscripto Codice Senensi nunc primùm editum, unà cum Notis Huberti Benvoglienti Patricii Senensis*, in L. Muratori (ed.), Rerum Italicarum scriptores ab anno Ærae Christianae quingentesimo ad millesimumquingentesimum ex Ambrosianae, Estensis, aliarumque insignium bibliothecarum codicibus … Tomus decimusquintus. Mediolani: ex typographia Societatis Palatinae in Regia Curia, 1729, pp. 1–128

Vetus Testamentum (Septuaginta): A. Rahlfs (ed.), *Septuaginta, Id Est Vetus Testamentum graece iuxta LXX interpretes*. Editio octava, 2 volumes. Stuttgart: Württembergische Bibelanstalt, 1965.

Villani, *Nuova Cronica*: G. Porta (ed.), *Giovanni Villani, Nuova Cronica*, 3 volumes. Parma: Fondazione Pietro Bembo/Ugo Guanda, 1990–1991.

Vita Sancti Symeonis Stylitae: P. van den Ven (ed.), *La vie ancienne de S. Syméon Stylite le jeune (521–592)*, 2 volumes (Subsidia hagiographia 32). Bruxelles: Société des Bollandistes, 1962–1970.

Vita Theodori Syceonis: A.-J. Festugière (ed.), *Vie de Théodore de Sykéon*. I. Texte grec. II. Traduction, commentaire et appendice, 2 volumes (Subsidia hagiographia 48). Bruxelles: Société des Bollandistes, 1970.

Secondary literature

Abbas 2012: A. Abbas, A. Lichtman, and S. Pillai, *Cellular and Molecular Immunology*. Philadelphia: Elsevier-Saunders, 2012.

Abu-Zeid 1983: M. Abu-Zeid, "The river Nile: main water transfer projects in Egypt and impacts on Egyptian Agriculture", in A. Biswas et al. (eds), *Long-Distance Water Transfer*. Dublin: United Nations University, 1983, pp. 6–34.

Achtman 1999: M. Achtman et al., "*Yersinia pestis*, the cause of plague, is a recently emerged clone of *Yersinia pseudotuberculosis*", *Proceedings of the National Academy of Science of the United States of America* 96 (24) (1999), pp. 14043–14048.

Achtman 2004: M. Achtman et al., "Microevolution and history of the plague bacillus, *Yersinia pestis*", *Proceedings of the National Academy of Science of the United States of America* 101 (51) (2004), pp. 17837–17842.

Adamson 2004: M. Adamson, *Food in Medieval Times*. London: Greenwood Press, 2004.

Adeolu 2016: M. Adeolu, S. Alnajar, S. Naushad, and R. Gupta, " Genome-based phylogeny and taxonomy of the 'Enterobacteriales': proposal for Enterobacterales ord. nov. divided into the families Enterobacteriaceae, Erwiniaceae fam. nov., Pectobacteriaceae fam. nov., Yersiniaceae fam. nov., Hafniaceae fam. nov., Morganellaceae fam. nov., and Budviciaceae fam. nov.", *International Journal of Systematic and Evolutionary Microbiology* 66 (2016), pp. 5575–5599.

Alberts 1998: B. Alberts et al., *Essential Cell Biology: an introduction to the Molecular Biology of the Cell*. New York, NY, and London: Garland Publishing, 1998.

Albrecht 1900: H. Albrecht, and A. Ghon, "Bacteriologische Untersuchungen über den Pest-bacillus", *Denkschriften Wissenschaft Wien* 66 (1900), pp. 1–353.

Al-Herz 2012: W. Al-Herz, and L. Notarangelo, "Classification of primary immunodeficiency disorders: one-fits-all does not help anymore", *Clinical Immunology* 144 (1) (2012), pp. 24–25

Allam 2002: K. Allam, A. Shalaby, and M. Ashour, "Seasonal distribution of fleas infesting rodents in various Egyptian eco-geographical areas and their susceptibility to malathion", *Journal of Egyptian Society of Parasitology* 32 (2) (2002), pp. 405–414.

Ally 1984: W. Ally, "The Palmer Drought Severity Index: limitations and assumptions", *Journal of Climate* 23 (1984), pp. 1100–1109.

Alonso 1999: J. Alonso, "Ecological interactions among *Yersinia pestis* in the common reservoir: the rodent", *Bulletin de la Société de Pathologie Exotique* 92 (1999), pp. 414–417.

American Hospital Formulary Service 2000: American Hospital Formulary Service, *Drug Information*. Bethesda, MD: American Society of Health System Pharmacists, 2000.

Ampel 1991: N. Ampel, "Plagues-What's past is present: Thoughts on the origin and history of new infectious diseases", *Reviews of Infectious Diseases* 13 (1991), pp. 658–656.

Andam 2016: C. Andam et al., "Microbial Genomics of ancient plagues and outbreaks", *Trends in Microbiology* 24 (12) (2016), pp. 978–990.

Anderson 1982: R. Anderson, "Directly transmitted viral and bacterial infections of man", in. R. Anderson (ed.), *The population dynamics of infectious diseases: theory and applications*. London: Chapman and Hall,1982, pp. 1–37.

Andréades 1920: A. Andréades, "De la population de Constantinople sous les empereurs byzantins", *Metron* 2 (1920), pp. 69–119.

Andrianaivoarimanana 2013: V. Andrianaivoarimanana et al., "Understanding the Persistence of Plague Foci in Madagascar", *PLoS Neglected Tropical Diseases* 7 (11) (2013), e2382.

Appleby 1980: A. Appleby, "The disappearance of plague: a continuing puzzle", *Economic History Review* 33 (2) (1980), pp. 161–173.

Asonitis 1987: Σ. Ασωνίτης, and Ι. Λασκαράτος, "Νεότερες ειδήσεις για την υγειονομική κατάσταση της Κέρκυρας κατά την πρώιμη Βενετοκρατία", *Δελτίον της Ιστορικής και Εθνολογικής Εταιρείας της Ελλάδος* 30 (1987), p. 1–18.

Atkinson 2016: S. Atkinson, and P. Williams, "Yersinia virulence factors—a sophisticated arsenal for combating host defences", *F1000 Research* 5 (2016, June 14), p. 1370.

Aufderheide 1998: A. Aufderheide, and C. Rodríguez-Martín, *The Cambridge Encyclopedia of Human Paleopathology*. Cambridge: Cambridge University Press, 1998.

Ayyadurai 2008: S. Ayyadurai et al., "Long-term persistence of virulent *Yersinia pestis* in soil", *Microbiology* 154 (2008), pp. 2865–2871.

Bacot 1924: A. Bacot, and C. Martin, "The respective influence of temperature and moisture upon the survival of the rat flea", *Journal of Hygiene* 23 (1924), pp. 98–105.

Bae 2014: J. Bae, "Meta-epidemiology", *Epidemiology and Health* 36 (2014), e2014019.

Baker 1984: R. Baker, "Commingling of Norway and Roof Rats with native rodents", in D. O. Clark, R. E. Marsh, and D. E. Beadle (eds), *Proceedings of the Eleventh Vertebrate Pest Conference (1984), March 6, 7, and 8, 1984, March 6, 7 and 8, 1984, Held at the Red LionMotor Inn, Sacramento, California*. University of California Press: Davis, CA, 1984, pp. 103–111.

Bales 1946: R. Bales, "Cultural differences in roles on alcohol", *Quarterly Journal of Studies Alcohol* 6 (1946), pp. 480–499.

Baltazar 1963: M. Baltazar et al., "The interepizootic preservation of plague in an inveterate focus. Working hypothesis", *Bulletin de la Société Pathologie Exotique* 56 (1963), pp. 1230–1245.

Baltazar 1964: M. Baltazar, "The conservation of plague in inveterate", *Journal of Hygiene, Epidemiology, Microbiology and Immunology* 120 (1964), pp. 409–421.

Bardet 1999: J. Bardet, and J. Dupâquier, *Histoire des populations de l'Europe, III. Les temps incertains: 1914–1998*. Paris: Fayard, 1999.

Barnes 1992: A. Barnes, and T. Quan, "*Plague*", in S. Gorbach et al. (eds), *Infectious Diseases*. Philadelphia, PA: W. B. Saunders, 1992, pp. 1285–1291.

Bartsocas 1966: C. Bartsocas, "Two Fourteenth Century Greek Descriptions of the Black Death", *Journal of the History of Medeicine* 21 (1966), pp. 394–400.

Begon 2006: M. Begon et al., "Epizootiologic parameters for plague in Kazakhstan", *Emerging Infectious Diseases* 12 (20) (2006), pp. 268–273.

Ben Ari 2008: T. Ben Ari et al., "Human plague in the USA: the importance of regional and local climate", *Biology Letters* 4 (2008), pp. 737–740.

Ben Ari 2010: T. Ben Ari et al., "Interannual variability of human plague occurrence in the Western United States explained by tropical and North Pacific Ocean climate variability", *American Journal of Tropical Medicine and Hygiene* 83 (2010), pp. 624–632.

Ben Ari 2011: T. Ben Ari et al., "Plague and Climate: Scales Matter", *PLoS Pathogens* 7 (9) (2011), e1002160.

Benedictow 2010: O. J. Benedictow, *What disease was plague? On the controversy over the Microbiological Identity of plague epidemics of the past.* Leiden, and Boston, MA: Brill, 2010.

Benedictow 2016: O. J. Benedictow, *The Black Death and later plague epidemics in the Scandinavian countries: Perspectives and controversies.* Warsaw and Berlin: De Gruyter 2016.

Bercovier 1984: H. Bercovier et al., "*Yersinia aldovae* (formerly *Yersinia enterocolitica*-like group X2): a new species of Enterobacteriaceae isolated from aquatic ecosystems", *International Journal of Systematic Bacteriology* 34 (1984), pp. 166–172.

Berdiner 1989: E. Berdiner, "Alexandre Yersin: pursues of plague", *Hospital Practice* 24 (1989), pp. 121–148.

Berheim 1978: F. Berheim, and A. Zenner, "The Sminthian Apollo and the epidemics among the Achaeans of Troy", *Transactions of the American Philololgical Association* 108 (1978), pp. 11–14.

Bertherat 2003: E. Bertherat et al., "Plague reappearance in Algeria after 50 years, 2003", *Emerging Infectious Diseases* 13 (10) (2003), pp. 1459–1462.

Bertherat 2005: E. Bertherat et al., "Epidémie de peste pulmonaire dans un camp minier de la République démocratique du Congo: le réveil brutal d'un vieux fléau", *Medicina Tropical* 65 (2005), pp. 511–514.

Beutler 1978: E. Beutler, *Hemolyticanemia in disorders of red cell metabolism.* New York, NY: Plenum Medical Books, 1978.

Bi 2012: Y. Bi et al., "*Yersinia pestis* versus *Yersinia pseudotuberculosis*: Effects on Host Macrophages", *Scandinavian Journal of Immunology* 76 (6) (2012), pp. 541–551.

Bibel 1976: D. Bibel, and T. Chen, "Diagnosis of plague: an analysis of the Yersin-Kitasato controversy", *Bacteriology Review* 40 (1976), pp. 633–651.

Biraben 1975: J.-N. Biraben, and J. Le Goff, "The plague in the early middle ages", in R. Forster, and O. Ranum (eds), *Biology of man in History.* Baltimore, MD: Johns Hopkins University Press, 1975, pp. 48–80.

Biraben 1989: J-N. Biraben, "Rapport: la peste du VIᵉ siècle dans l'Empire byzantin", in J. Le Fort, C. Morrisson, and J.-P. Sodini (eds), *Hommes et richesses dans l'Empire byzantin. IVᵉ–VIIᵉ siècles.* Paris: Lethielleux, 1989, pp. 121–125.

Bitam 2006: I. Bitam et al., "Zoonotic focus of Plague, Algeria", *Emerging Infectious Diseases* 12 (12) (2006), pp. 1975–1977.

Bobrov 1997: A. Bobrov, and A. Filipov, "Prevalence of IS285 and IS100 in *Yersinia pestis* and *Yersinia pseudotuberculosis* genomes", *Molekuliarnaia Genetika, Mikrobiologiia i Virusologiia* 2 (1997), pp. 36–40.

Böhme 2012: K. Böhme et al., "Concerted actions of a thermo-labile regulator and a unique intergenic RNA thermosensor control Yersinia virulence", *PLoS Pathogens* 28 (2) (2012), e1002518.

Bos 2011: K. Bos et al., "A draft genome of *Yersinia pestis* from victims of the Black Death", *Nature* 478 (7370) (2011), pp. 506–510.

Bos 2015: K. Bos et al., "Parallel detection of ancient pathogens via array-based DNA capture", *Philosophical Transactions of the Royal Society of London*, B: *Biological Sciences* 370 (1660) (2015), 20130375.

Bos 2016: K. Bos et al., "Eighteenth century *Yersinia pestis* genomes reveal the long-term persistence of an historical plague focus", *bioRxiv* 5 (2016), e12994.

Bouffard 1923: G. Bouffard, and G. Girard, " Le dépistage de la peste par ponction du foie", *Bulletin de la Société de Pathologie Exotique* 16 (1923), pp. 501–524.

Bouras 1981: C. Bouras, "City and Village: Urban Design and Architecture", in H. Hunger, W. Hörander, C. Cupane, E. Kislinger, and J. Raasted (eds), *XVI. Internationaler Byzantinistenkongress, Wien 4.–9. Oktober 1981. Akten*, I. Teil: Hauptreferate, 2 volumes (*Jahrbuch der Österreichischen Byzantinistik* 31). Wien: Österreichische Akademie der Wissenschaften, 1981, vol. 2, pp. 611–653.

Bouras 2012: K. Μπούρας, "Μεσοβυζαντινές και υστεροβυζαντινές πόλεις από τη σκοπιά της πολεοδομίας και της αρχιτεκτονικής", in Τ. Κιουσοπούλου (ed.), *Οι Βυζαντινές πόλεις (8ος-15οςαιώνας): προοπτικές της έρευνας και νέες ερμηνευτικές προσεγγίσεις*. Ρέθυμνο: Πανεπιστημιακές Εκδόσεις Κρήτης, 2012, pp. 1–14.

Bowsky 1964: W. Bowsky, "The impact of the Black Death upon Sienese government and society", *Speculum* 39 (1) (1964), pp. 1–34.

Braton 1981: T. Braton, "The identity of the plague of Justinian-PartI", *Transactions and Studies of the Gollege of Physicians of Philadelphia* 3 (3) (1981), pp. 113–124.

Brooks 1998: G. Brooks, *Medical Microbiology*. Stamford, CT: Appleton & Lange, 1998.

Brown 2002: J. Brown, and S. Ernest, "Rain and rodents: Complex dynamics of desert consumers", *Bioscience* 52 (2002), pp. 979–987.

Brown 2008: T. Brown, "Byzantine Italy (680–876)", in J. Shepard (ed.), *The Cambridge History of the Byzantine Empire*. Cambridge: Cambridge University Press, 2008, pp. 433–464.

Browning 1978: R. Browning, " Ὁ αἰώνας τοῦ Ἰουστινιανοῦ", in Γ. Χριστόπουλος, Ἰ. Μπαστιᾶς (eds), *Ἱστορία τοῦ Ἑλληνικοῦ Ἔθνους. Τόμος Ζ´: Βυζαντινὸς Πολιτισμός. Πρωτοβυζαντινοὶ Χρόνοι*. Ἀθῆνα: Ἐκδοτικὴ Ἀθηνῶν, 1978, pp. 150–204.

Brubaker 1972: R. Brubaker, "The genus Yersinia: biochemistry and genetics of virulence", *Current Topics in Microbiology and Immunology* 57 (1972), pp. 111–158.

Brubaker 2003: R. Brubaker, "Interleukin-10 and Inhibition of Innate Immunity to Yersiniae: Roles of Yops and LcrV (V Antigen)", *Infection and Immunity* 71 (7) (2003), pp. 3673–3681.

Brygoo 1965: E. Brygoo, and A. Dodin, "A propos de la peste tellurique et de la peste fouissement. Données malgaches", *Bulletin de la Société de Pathologie Exotique* 1 (1965), pp. 14–17.

Bubeck 2007: S. Bubeck, A. Cantwell, and P. Dube, "Delayed inflammatory response to primary pneumonic plague occurs in both outbred and inbred mice", *Infection and Immunity* 75 (2007), pp. 697–705.

Buchon 1845: J. Buchon, *Nouvelles recherches historiques sur la Principauté de Morée*. Paris: J. Renouard, 1845.

Büntgen 2012: U. Büntgen et al., "Digitizing historical plague", *Clinical Infectious Diseases* 55 (11) (2012), pp. 1586–1588.

Burlington 1967: R. Burlington, and G. Klain, "Gluconeogenesis during hibernation and arousal from hibernation", *Comparative Biochemistry and Physiology* 22 (1967), pp. 701–708.

Burrougs 1947: A. Burrougs, "Sylvatic plague studies. The vector efficiency of nine species of fleas compared with *Xenopsylla cheopis*", *Journal of Hygiene* 45 (1947), pp. 371–396.

Burrows 1964: T. Burrows, J. Farrell, and W. Gillet, "The catalase activities of *Pasturela pestis* and other bacteria", *British Journal of Experimental Pathology* 45 (1964), pp. 579–588.

Califf 2015: K. Califf et al., "Redefining the differences in gene content between *Yersinia pestis* and *Yersinia pseudotuberculosis* using large-scale comparative genomics", *Microbial Genome* 1(2) (2015), e000028.

Camacho 2018: A. Camacho et al., "Cholera epidemic in Yemen, 2016–2018: an analysis of surveillance data", *Lancet Global Health* 6 (6) (2018), e680-e690.

Campbell 1998: G. Campbell, and D. Dennis, "Plague and others Yersinia infections", in D. Kasper et al. (eds), *Harrison's Principles of Internal Medicine*. New York, NY: McGraw Hill, 1998, pp. 975–983.

Cantor 2001: N. Cantor, *The Black Death and the World it made*. London: Simon and Schuster, 2001.

Carey 2003: H. Carey, M. Andrews, and S. Martin, "Mammalian Hibernation: Cellular and Molecular Responses to Depressed Metabolism and Low Temperature", *Physiological Reviews* 83 (2003), pp. 1153–1181.

Carmichael 2014: A. G. Carmichael, "Plague Persistence in Western Europe. A Hypothesis", *The Medieval Globe* 1 (1) (2014), pp. 157–191.

Carniel 2001: E. Carniel, "The Yersinia high-pathogenicity island: an iron-uptake island", *Microbes and Infections* 3 (2001), pp. 561–569.

Carr 2015: M. Carr, *Merchant Crusaders in the Aegean, 1291–1352*. Rochester, NY: Boydell Press, 2015.

Cartwright 1972: F. Cartwright, *Disease and History*. New York, NY: Barnes & Nobles, 1972.

Casson 1989: L. Casson, *The Periplus Maris Erythraei*. Text, translation and commentary. Princeton, NJ: Princeton University Press, 1989.

Cavanaugh 1968: D. Cavanaugh et al., "Some observations on the current plague outbreak in the Republic of Vietnam", *American Journal of Public Health and Nation's Health* 58 (1968), pp. 742–752.

Cavanaugh 1971: D. Cavanaugh, "Specific effect of temperature upon transmission of the plague bacillus by the oriental rat flea, *Xenopsylla cheopis*", *American Journal of Tropical Medicine and Hygiene* 20 (1971), pp. 264–272.

Cavanaugh 1972: D. Cavanaugh, and J. Marshall, "The influence of climate on the seasonal prevalence of plague in the Republic of Vietnam", *Journal of the Wildlife Diseases* 8 (1972), pp. 85–94.

Centers for Disease Control and Prevention 1982: Centers for Disease Control and Prevention, "Plague vaccine. Selected recommendations of the Public Health Service Advisory Committee on Immunization Practices (ACIP)", *Morbidity and Mortality Weekly Report* 40 (RR-21) (1982), pp. 41–42.

Centers for Disease Control and Prevention 2001: Centers for Disease Control and Prevention, *Facts about pneumonic plague. Emergency Preparedness and Response*. Atlanta, GA: U.S. Department of Health and Human Services/CDC, 14 October 2001.

Centers for Disease Control and Prevention 2012: Centers for Disease Control and Prevention, *Principles of Epidemiology in Public Health Practice. An Introduction to Applied Epidemiology and Biostatistics*. Atlanta, GA: U.S. Department of Health and Human Services/ Centers for Disease Control and Prevention CDC, 2012.

Cerdeno-Tarraga 2004: A. Cerdeno-Tarraga, N. Thomson, and J. Parkhill, "Pathogens in decay", *Nature Reviews Microbiology* 2 (10) (2004), pp. 774–775.

Chain 2004: P. Chain et al., "Insights into the evolution of *Yersinia pestis* through whole-genome comparison with *Yersinia pseudotuberculosis*", *Proceedings of the National Academy of Sciences of the United States of America* 101 (38) (2004), pp. 13826–13831.

Chain 2006: P. Chain et al., "Complete Genome Sequence of *Yersinia pestis* Strains Antiqua and Nepal516: Evidence of Gene Reduction in an Emerging Pathogen", *Journal of Bacteriology* 188 (12) (2006), pp. 4453–4463.

Chappell 1971: J. Chappell, "Climatic pulsations in inner Asia and correlations between sunspots and weather", *Paleogeography, Paleoclimatology and Paleoecology* 10 (1971), pp. 177–197.

Charanis 1967: P. Charanis, "Observations on the demography of the Byzantine Empire", in J. N. Hussey, D. Obolensky, and S. Runciman (eds), *Proceedings of the XIIIth International Congress* *of Byzantine studies. Oxford 5–10 September 1966*. Oxford: Oxford University Press, 1967, pp. 445–463.

Che 2010: D. Che et al., "Classification of genomic islands using decision trees and their ensemble algorithms", *Genomics* 11 (Suppl 2) (2010), S1.1471–2164/11/S2/S1.

Chen-Chia 1997: C. Chen-Chia, "Diphtheria", in J. Johnson, and Y.Victor (eds), *Infectious Diseases and Antimicrobial Therapy of the Ears, Nose and Throat*. Philadelphia, PA: W. B. Saunders, 1997, pp. 453–458.

Chircop 2018: J. Chircop, "Quarantine, sanitization, colonialism and the construction of the "contagious Arab" in the Mediterranean, 1830s-1900", in J. Chircop, and F. Martinez (eds), *Mediterranean quarantines, 1750–1914: Space, identity and power*. Manchester: Manchester University Press, 2018, pp. 199–231.

Christie 1980: A. Christie, T. Chen, and S. Elberg, "Plague in camels and goats: their role in human epidemics", *Journal of Infectious Diseases* 141 (1980), pp. 724–726.

Chunha 2008: C. B. Cunha, and B. A. Cunha, "Great plagues of the past and remaining questions", in D. Raoult and M. Drancourt (eds), *Paleomicrobiology-Past human infections*. Berlin and Heidelberg: Springer, 2008, pp. 4–9.

Clavijo 1859: R. Clavijo, *Narrative of the embassy to the court of Timour at Samarcand A.D. 1403–1406*. London: Hakluyt Society, 1859.

Coburn 2007: B. Coburn, I. Sekirov, and B. Finlay, "Type III secretion systems and disease", *Clinical Microbiology Review* 20 (4) (2007), pp. 535–549.

Cohn 2002: S. Cohn, "The Black Death: End of a Paradigm", *American Historical Review* 107(3) (2002), pp. 703–738.

Cohn 2007: S. Cohn, "After the Black Death: labour legislation and attitudes towards labour in late-medieval western Europe", *Economic History Review* 60 (3) (2007), pp. 457–485.

Cohn 2008: S. Cohn, "Epidemiology of the Black Death and Successive Waves of Plague", *Medical History*, Supplement 27 (2008), pp. 74–100.

Colin 1981: J. Colin, *Cyriaque d'Ancône. Humaniste, grand voyageur et fondateur de la science archéologique*. Paris: Maloine, 1981.

Collin 2007: A. Collin, *Land Transport in Roman Egypt. A Study of Economics and Administration in a Roman Province* (Oxford Classical Monographs). Oxford: Oxford University Press, 2007.

Committee of Infectious Diseases 1997: Committee of Infectious Diseases, "Plague", in G. Peter (ed.), *Redbook*. Elk Grove, IL: American Academy of Pediatrics, 1997, pp. 408–410.

Concina 2003: E. Concina, *La città Bizantina*. Roma and Bari: Laterza, 2003.

Congourdeau 1993: M.-H. Congourdeau, "La société Byzantine face aux grandes pandémies", in E. Patlagean (ed.), *Maladie et Société à Byzance* (Collectanea 3). Spoleto: Centro Italiano di Studi sull'Alto Medioevo, 1993, pp. 21–42.

Congourdeau 1999: M.-H. Congourdeau, "La peste noire à Constantinople de 1348 à 1466", *Medicina nei Secoli* 11 (2) (1999), pp. 377–389.

Conrad 1981: L. Conrad, "Arabic plague chronologies and treatises: social and historical factors in the formation of a literary genre". *Studia Islamica* 54 (1981), pp. 51–93.

Conrad 1982: L. Conrad, "Ta'ūnand Wabā: Conceptions of plague and pestilence in Early Islam", *Journal of Economic and Social History of the Orient* 25 (1982), pp. 268–307.

Conrad 1994: L. Conrad, "Epidemic disease in central Syria in the late sixth century. Some new insights from the verse of Hassān ibn Thābit", *Byzantine and Modern Greek Studies* 18 (1994), pp. 17–26.

Cook 1998: R. Cook, "Alcohol abuse, alcoholism, and damage to the immune system-a review", *Alcoholism, Clinical and Experimental Researsch* 22 (1998), pp. 1927–1942.

Cook 2015: E. Cook et al., "Old World megadroughts and pluvial during the Common Era", *Science Advances* 1 (10) (2015), pp. 1–9.

Cooper 2000: A. Cooper, and H. Poinar, "Ancient DNA: do it right or not at all", *Science* 289 (5482) (2000), p. 1139.

Cooper 2006: M. Cooper, and M. Alder, "The Evolution of Adaptive Immune Systems", *Cell* 124 (2006), pp. 815–822.

Cornelis 2002: G. Cornelis, "Yersinia type III secretion: send in the effectors", *Journal of Cellular Biology* 158 (2002), pp. 401:408.

Couper 2008: K. Couper, D. Blount, and E. Riley, "IL-10: The master regulator of immunity to infection", *Journal of Immunology* 180 (2008), pp. 5771–5777.

Crespo 2014: F. Crespo, and M. B. Lawrenz, "Heterogeneous immunological landscapes and medieval plague: an invitation to a new dialogue between historians and immunologists", *The Medieval Globe* 1 (2014), pp. 229–257.

Cui 2013: Y. Cui et al., "Historical variations in mutation rate in an epidemic pathogen, *Yersinia pestis*", *Proceedings of the National Academy of Science of the United States of America* 110 (2) (2013), pp. 577–582.

Cui 2016: Y. Cui, and Y. Song, "Genome and Evolution of *Yersinia pestis*", in R. Yang, and A. Anisimov (eds), *Yersinia pestis: Retrospectives and Perspectives* (Advances in Experimental Medicine and Biology 918). New York, NY: Springer, 2016, pp. 171–192.

Cuntz 1929: O. Cuntz, *Itineraria Romana*. Leipzig: Teubner, 1929.

Curtin 1908: J. Curtin, *The Mongols: A History*. Boston, MA: Little-Brown & Co, 1908.

Dagron 1974: G. Dagron, *Naissance d'une capitale: Constantinople et ses institutions de 330 à 451* (Bibliothèque Byzantine, Études, 7). Paris: Presses Universitaires de France, 1974.

Dagron 1984: G. Dagron, "Les villes dans l'Illyricum protobyzantin", in R. Manuel (ed.), *Villes et peuplement dans l'Illyricum protobyzantin. Actes du colloque de Rome (12–14 mai 1982)* (Publications de l'École française de Rome 77). Rome: École française de Rome, 1984, pp. 1–20.

Dai 2005: E. Dai et al., "Identification of different regions among strains of *Yersinia pestis* by suppression subtractive hybridization", *Research in Microbiology* 156 (7) (2005), pp. 785–789.

Darby 2002: C. Darby et al., "*Caenorhabditis elegans*: plague bacteria biofilms block food intake", *Nature* 417 (6886) (2002), pp. 243–244.

Davidson 2018: S. Davidson, "Treating influenza infection, from now and into the future", *Frontiers in Immunology* 9 (2018), p. 1946.

Davis 1953: D. Davis, "Plague in Africa from 1935 to 1949: a survey of wild rodents in African territories", *Bulletin ofthe World Health Organization* 5 (1953), pp. 665–700.

Davis 2005: S. Davis, E. Calvet, and H. Leirs, "Fluctuating rodent populations and risk to humans from rodent-borne zoonoses", *Vector Born Zoonotic Diseases* 5 (2005), pp. 305–314.

De Vaan 2002: M. De Vaan, *Etymological Dictionary of Latin*. Leiden: Brill, 2002.

Deng 2002: W. Deng et al., "Genome sequence of *Yersinia pestis* KIM", *Journal of Bacteriology* 184 (16) (2002), pp. 4601–4611.

Dennis 2000: D. Dennis, "Plague", in G. Strickland (ed.), *Hunter's Tropical Medicine and Emerging Infectious Diseases*. Philadelphia, PA: W. B. Saunders, 2000, pp. 402–411.

Derbes 1966: V. Derbes, "De Mussis and the great plague of 1348", *Journal of the American Medical Association* 196 (1966), pp. 179–182.

Detorakis 1970–71: Th. Detorakis, "Η πανώλης εν Κρήτη. Συμβολή εις την ιστορίαν των επιδημιών της νήσου ", *Επιστημονική Επετηρίς της Φιλοσοφικής Σχολής του Πανεπιστημίου Αθηνών* 21 (1970–1971), pp. 118–136.

Devault 2014: A. Devault et al., "Ancient pathogen DNA in archaeological samples detected with a Microbial Detection Array", *Scientific Reports* 4 (2014), pp. 42–45.

Devignat 1951: R. Devignat, "Varieties of *Pasteurella pestis*; new hypothesis", *Bulletin of the World Health Organization* 4 (1951), pp. 247–263.

Didelot 2017: X. Didelot, L. Whittles, and I. Hall, "Model-based analysis of an outbreak of bubonic plague in Cairo in 1801", *Journal of the Royal Society of Interface* 14 (131) (2017), 20170160.

Dietz 1993: K. Dietz, "The estimation of the basic reproduction number for infectious diseases", *Statistical Methods in Medical Research* 2 (1993), pp. 23–41.

Dietzel 2017: E. Dietzel et al., "Functional Characterization of Adaptive Mutations during the West African Ebola Virus Outbreak", *Journal of Virology* 91 (2) (2017), e01913–16.

Dirckx 1985: J. Dirckx, "The Biblical plague of "hemorrhoids": An outbreak of bilharziasis", *American Journal of Dermatopathology* 7 (4) (1985), pp. 341–346.

Djurdjevic 2009: P. Djurdjevic et al., "Role of decreased production of interleukin-10 and interferon-gamma in spontaneous apoptosis of B-chronic lymphocytic leukemia lymphocytes *in vitro*", *Archives of Medical Research* 40 (2009), pp. 357–363.

Dolls 1977: M. Dolls, *The Black Death in the Middle East*. Princeton, NJ: Princeton Univerity Press, 1977.

Dongsheng 2014: D. Dongsheng et al., "Identifying Pathogenicity Islands in Bacterial Pathogenomics Using Computational Approaches", *Pathogens* 3 (2004), pp. 36–56.

Donoghue 2008: H. Donoghue, "Palaeomicrobiology of Tuberculosis", in D. Raoult, and M. Drancourt (eds), *Paleomicrobiology-Past Human Infections*. Berlin and Heidelberg: Springer, 2008, pp. 75–97.

Dooley 2007: A. Dooley, "The plague and its consequences in Ireland", in L. Little (ed.), *Plague and the End of Antiquity. The Pandemic 541–750*. Cambridge and New York, NY: Cambridge University Press and The American Academy in Rome, 2007, pp. 215–228.

Drancourt 1998: M. Drancourt et al., "Detection of 400-year-old *Yersinia pestis* DNA in human dental pulp: an approach to the diagnosis of ancient septicemia", *Proceedings of the National Academy of Science of the United States of America* 95 (21) (1998), pp. 12637–12640.

Drancourt 2002: M. Drancourt, and D. Raoult, "Molecular insights into the history of plague", *Microbes and Infections* 4 (2002), pp. 105–109.

Drancourt 2004: M. Drancourt et al., "Genotyping, Orientalis-like *Yersinia pestis*, and plague pandemics", *Emerging Infectious Diseases* 10 (9) (2004), pp. 1585–1592.

Drancourt 2006: M. Drancourt, L. Houhamdi, and D. Raoult, "*Yersinia pestis* as a telluric, human ectoparasite-borne organism", *Lancet Infectious Diseases* 6 (4) (2006), pp. 234–241.

Drancourt 2008: M. Drancourt, and D. Raoult, "Molecular Detection of Past Pathogens", in D. Raoult, and M. Drancourt (eds), *Paleomicrobiology-Past Human Infections*. Berlin and Heidelberg: Springer, 2008, pp. 55–68.

Drancourt 2012: M. Drancourt, "Plague in the genomic area", *Clinical Microbiology and Infection* 18 (2012), pp. 224–230.

Drancourt and Raoult 2016: M. Drancourt, and D. Raoult, "Molecular history of plague", *Clinical Microbiology and Infection* 22 (11) (2016), pp. 911–915.

Du 2002: Y. Du, R. Rosqvist, and A. Forsberg, "Role of fraction 1 antigen of *Yersinia pestis* in inhibition of phagocytosis", *Infection and Immunity* 70 (2002), pp. 1453–1460.

Duchemin 2015: W. Duchemin et al., "Reconstruction of an ancestral *Yersinia pestis* genome and comparison with an ancient sequence", *Biomed Central Genomics* 16 (2015), S9.

Duncan-Jones 1996: R. Duncan-Jones, "The impact of the Antonine Plague", *Journal of Roman Archaeology* 9 (1996), pp. 108–136.

Dunning 2011: H. Dunning, "Horizontal gene transfer between bacteria and animals", *Trends in Genetics* 27 (4) (2011), pp. 157–163.

Duplantier 2005: J. Duplantier et al., "From the recent lessons of the Malagasy foci towards a global understanding of the factors involved in plague re-emergence", *Veterinary Research* 36 (2005), pp. 437–453.

Eadie 1982: J. Eadie, "City and countryside in Late Roman Pannonia", in R. Hohlfedes (ed.), *City, Town and countryside in the Early Byzantine Era*. New York, NY: Columbia University Press, 1982, pp. 25–41.

Eads 2016: D. Eads et al., "Droughts may increase susceptibility of prairie dogs to fleas: incongruity with hypothesized mechanisms of plague cycles in rodents", *Journal of Mammalogy* 97 (4) (2016), pp. 1044–1053.

Economidis 1978: N. Οἰκονομίδης, " Ἡ ἑνοποίηση τοῦ Εὐρασιατικοῦ χώρου (945–1071)", in Γ. Χριστόπουλος, Ἰ. Μπαστιᾶς (eds), *Ἱστορία τοῦ Ἑλληνικοῦ Ἔθνους*. Τόμος Ζ΄: Βυζαντινὸς Πολιτισμός. Πρωτοβυζαντινοὶ Χρόνοι. Ἀθήνα: Ἐκδοτικὴ Ἀθηνῶν, 1978, pp. 98–148.

Efthychiades 1983: A. Efthychiades, *Εἰσαγωγή εἰς τήν Βυζαντινήν Θεραπευτικήν*. Ἀθήνα: Παρισιάνος, 1983.

Eidson 1988: M. Eidson et al., "Feline plague in New Mexico: risk factors and transmission to humans", *American Journal of Public Health* 78 (1988), pp. 1333–1335.

Eisen 2008: R. Eisen et al., "Persistence of *Yersinia pestis* in soil under natural conditions", *Emerging Infectious Diseases* 14 (6) (2008), pp. 941–943.

Eitzen 1997: E. Eitzen, and E. Takafuji, "Historical Overview of Biological Warfare", in F. Sidell, E. Takafuji, and D. Franz (eds), *Medical aspects of Chemical and Biological Warfare* (Textbook of Military Medicine Series). Houston, TX: Office of The Surgeon General, Borden Institute, US Army Medical Department Center and School Health Readiness, Center of Excellence Fort Sam, 1997, pp. 415–423.

El-Kady 1998: G. El-Kady et al., "Rodents, their seasonal activity, ecto- and blood-parasites in Saint Catherine area, South Sinai Governorate", *Journal of the Egyptian Society of Parasitology* 28 (3) (1998), pp. 815–826.

Ell 1987: S. Ell, "Plague and leprosy in the Middle Ages: a paradoxical cross-immunity?", *International Journal of Leprosy and Other Mycobacterial Diseases* 55 (2) (1987), pp. 345–350.

Eltahir 1996: E. Eltahir, "El Niño and the Natural Variability in the Flow of the Nile River", *Water Resources Research* 32 (1996), pp. 131–137.

Emery 1967: R. Emery, "The Black Death of 1348 in Perpignan", *Speculum* 52 (4) (1967), pp. 611–623.

Enscore 2002: R. Enscore et al., "Modeling relationships between climate and the frequency of human plague cases in the southwestern United States, 1960–1997", *American Journal of Tropical Medicine and Hygiene* 66 (2002), pp. 186–196.

Eppinger 2010: M. Eppinger et al., "Genome Sequence of the Deep-Rooted *Yersinia pestis* Strain Angola Reveals New Insights into the Evolution and Pangenome of the Plague Bacterium", *Journal of Bacteriology* 192 (6) (2010), pp. 1685–1699.

Eroshenko 2017: G. Eroshenko et al., "*Yersinia pestis* strains of ancient phylogenetic branch 0.ANT are widely spread in the highmountain plague foci of Kyrgyzstan", *PLoS ONE* 12 (10) (2017), e0187230.

Fears 2004: J. Fears, "The plague under Marcus Aurelius and the decline and fall of the Roman Empire", *Infectious Diseases Clinics of North America* 18 (2004), pp. 65–77.

Feodorova 2000: V. Feodorova, and Z. Devdariani, "The interaction of *Yersinia pestis* with erythrocytes", *Journal of Medical Microbiology* 51 (2000), pp. 150–158.

Flajnik 2004: M. Flajnik, and L. Pasquier, "Evolution of innate and adaptive immunity: can we draw a line?", *Trends in Immunology* 25 (12) (2004), pp. 640–644.

Flajnik 2010: M. Flajnik, and M. Kasahara, "Origin and evolution of the adaptive immune system: genetic events and selective pressures", *Nature Reviews Genetics* 11 (1) (2010), pp. 47–59.

Fordham 1976: C. Fordham et al., "Bubonic plague from exposure to a rabbit: a documented case and a review of rabbit-associated plague in the United States", *American Journal of Epidemiology* 104 (1) (1976), pp. 81–87.

Fraedrich 1991: K. Fraedrich, and C. Bantzer, "A note to fluctuations of the Nile River Flood Levels (715–1470)", *Theoretical and Applied Climatology* 44 (1991), pp. 167–171.

Frederiksen 1964: W. Frederiksen, "A study of some *Yersinia pseudotuberculosis*-like (*Bacterium enterocoliticum* and *Pasteurella X*)", in *Proceedings of the 14th Scandinavian Congress of Pathology and Microbiology, Oslo, 1964*. Oslo: Universitetsforlaget, 1964, pp. 103–104.

Fukushima 2001: H. Fukushima et al., "*Yersinia enterocolitica* 09 as a possible barrier against *Yersinia pestis* in natural plague foci Ningxia, China", *Current Microbiology* 42 (2001), pp. 1–7.

Gage 2008: K. Gage et al., "Climate and vector-borne diseases", *American Journal of Preventive Medicine* 35 (2008), pp. 436–450.

Gal-Mor 2006: O. Gal-Mor, and B. B. Finlay, "Pathogenicity islands: a molecular toolbox for bacterial virulence", *Cellular Microbiology* 8 (11) (2006), pp. 1707–1719

Gani 2004: R. Gani, and S. Leach, "Epidemiologic determinants for modeling pneumonic plague outbreaks", *Emerging Infectious Diseases* 10 (4) (2004), pp. 608–614

Garcia 2007: E. Garcia et al., "Pestoides F, an atypical *Yersinia pestis* strain from the former Soviet Union", *Advances in Experimental Medicine and Biology* 603 (2007), pp. 17–22.

Garmichael 2008: A. Garmichael, "Universal and Particular: The language of plague 1348–1500", *Medical History*, Supplement 27 (2008), pp.17–52.

Girard 1951: G. Girard, "Reactions of leprous rats (Stefansky bacillus disease) to experimental plague infection", *Séances de la Société de Biologie et Filiales* 145 (21–22) (1951), pp. 1627–1630.

Girard 1975: G. Girard, "What was the fate of patients with leprosy during the plague pandemic in the Middle Ages (1348–1350)?", *Bulletin de la Société de Pathologie Exotique* 68 (1) (1975), pp. 33–37.

Glanville 1955: D. Glanville, "Earthquakes at Constantinople and Vicinity A.D. 342–1454", *Speculum* 30 (4) (1955), pp. 596–600.

Glucker 1987: C. A. M. Glucker, *The city of Gaza in the Roman and Byzantine Periods* (BAR International Series 325). Oxford: British Archaeological Reports, 1987.

Gophna 2003: U. Gophna, E. Rona, and D. Graurb, "Bacterial type III secretion systems are ancient and evolved by multiple horizontal-transfer events", *Gene* 312 (2003), pp. 151–156.

Gordon 1947: J. Gordon, and P. Knies, "Flea versus rat control in human plague", *American Journal of Medical Sciences* 213 (1947), pp. 362–367.

Gottfriend 1983: R. Gottfriend, *The Black Death*. London: MacMillan, 1983.

Gould 2009: D. Gould, "Isolation precautions to prevent the spread of contagious diseases", *Nursing Standard* 23 (22) (2009), pp. 47–55.

Gratz 1983: N. G. Gratz, and A. W. A. Brown, *Fleas-Biology and Control*. Geneva: World Health Organization, 1983 (unpublished document WHO/VBC/83.874).

Green 1978: C. Green, D. Gordon, and N. Lyons, "Biological species in *Praomys* (Mastomys) *natalensis* (Smith), a rodent carrier of Lassa virus and bubonic plague in Africa", *American Journal of Tropical Medicine and Hygiene* 27 (3) (1978), pp. 627–629.

Green 1984: C. Green et al., "Effect of hibernation on liver and kidney metabolism in 13-lined ground squirrels", *Comparative Biochemistry and Physiology* 79 (1984), pp. 167–171.

Gregory 1982: T. Gregory, "Fortification and Urban Design in Early Byzantine Greece", in R. Hohlfedes (ed.), *City, Town and Countryside in the Early Byzantine Era*. New York, NY: Columbia University Press, 1982, pp. 43–64.

Griffin 2011: J. Griffin et al., "Joint estimation of the basic reproduction number and generation time parameters for infectious disease outbreaks", *Biostatistics* 12 (2) (2011), pp. 303–312.

Gu 2007: J. Gu et al., "Genome evolution and functional divergence in *Yersinia*", *Journal of Experimental Zoology* 308B (2007), pp. 37–49.

Guillaume 2018: Y. Guillaume et al., "Responding to Cholera in Haiti: Implications for the National Plan to Eliminate Cholera by 2022", *Journal of Infectious Diseases* 218 (Suppl. 3) (2018), pp. S167–S170.

Guiyoule 1994: A. Guiyoule et al., "Plague pandemics investigated by ribotyping of *Yersinia pestis* strains", *Journal of Clinical Microbiology* 32 (3) (1994), pp. 634–641.

Gwilt 1986: J. Gwilt, "Biblical ills and remedies", *Journal of the Royal Society of Medicine* 79 (1986), pp. 738–741.

Gyles 2014: C. Gyles, and P. Boerlin, "Horizontaly transferred genetic elements and their role in pathogenesis of bacterial disease", *Veterinary Pathology* 51 (2) (2014), pp. 328–340.

Haas 2006: C. Haas, "The Antonine plague", *Bulletin de l'Académie nationale de Médecine* 190 (4–5) (2006), pp. 1093–1098.

Hacker 2000: J. Hacker, and J. Kaper, "Pathogenicity islands and the evolution of microbes", *Annual Review of Microbiology* 54 (2000), pp. 641–679.

Haensch 2010: S. Haensch et al., "Distinct Clones of *Yersinia pestis* caused the Black Death", *PLoS Pathogens* 6 (10) (2010), e1001134.

Haeser 1882: H. Haeser, *Lehrbuch der Geschichte der Medizin und der Epidemischen Krankheiten*. Jena: Gustav Fischer, 1882.

Halloran 2001: E. Halloran, "Concepts of Transmission and Dynamics", in J. Thomas, and D. Weber (eds), *Epidemiologic methods for the study of Infectious Diseases*. Oxford: Oxford University Press, 2001, pp. 56–86.

Harbeck 2013: M. Harbeck et al., "*Yersinia pestis* DNA from Skeletal Remains from the 6[th]Century AD Reveals Insights into Justinianic Plague", *PLoS Pathogens* 9 (5) (2013), e1003349.

Harper 2015: K. Harper, "Pandemics and passages to late antiquity: Rethinking the plague of c. 249–270 described by Cyprian", *Journal of Roman Archaeology,* 28 (2015), pp. 223–260.

Harris 2010: S. Harris et al., "Evolution of MRSA during hospital transmission and 5964) intercontinental spread", *Science* 327 (4964) (2010), pp. 469–474.

Hays 2005: J. Hays, *Epidemics and Pandemics: Their impacts on Human History*. Oxford: ABC Clio, 2005.

Hendrickx 2012: B. Hendrickx, "The border troops of the Roman-Byzantine southern Egyptian *limes:* problems and remarks on the role of the African and "black" African military units", Ἐκκλησιαστικὸς Φάρος 94 (23) (2012), pp. 95–114.

Herlihy 1997: D. Herlihy, *The Black Death and the transformation of the West*. Cambridge, MA: Harvard University Press, 1997.

Hethcote 1980: H. Hethcote, and D. Tudor, "Integral equation models for endemic infectious diseases", *Journal of Mathematical Biology* 9 (1980), pp. 37–47.

Hinckley 2012: A. Hinckley et al., "Transmission dynamics of primary pneumonic plague in the USA", *Epidemiology and Infection* 140 (3) (2012), pp. 554–560.

Ho 2009: S. Ho et al., "The association of virulence factors with genomic islands", *PLoS ONE* 4 (12) (2009), e8094.

Hoen 2015: A. Hoen et al., "Epidemic wave dynamics attributable to urban community structure: a theoretical characterization of disease transmission in a large network", *Journal of Medical Internet Research* 17 (7) (2015), e169.

Hunger 1976: H. Hunger, "Thukydides bei Johannes Kantakuzenos. Beobachtungen zur Mimesis", *Jahrbuch der Österreichischen Byzantinistik* 25 (1976), pp. 185–188.

Hussein 1955: A. Hussein, "Changes in the epidemiology of plague in Egypt 1899–1951", *Bulletin of the World Health Organization* 13 (1955), pp. 27–48.

Hymes 2014: R. Hymes, "Epilogue: A Hypothesis on the East Asian Beginnings of the *Yersinia pestis* Polytomy", *The Medieval Globe* 1 (1) (2014), pp. 289–293, 299.

Inglesby 2000: T. Inglesby et al., "Plague as a biological weapon: medical and public health management, Working Group on Civilian Biodefence", *Journal of the American Medical Association* 283 (17) (2000), pp. 2281–2290.

Ivanov 2013: O. Ivanov, "Medieval ports on the Southern Coast of the Crimean Peninsula: Navigation and Urbanisation", in F. Karagiani (ed.), *Proceedings of International Symposium "Medieval Ports in North Aegean and the Black Sea: Links to the Maritime Routes of the East" in the frame of the project "OLKAS: From Aegean to the Black Sea Medieval ports in the Maritime Routes of the East"*. Thessalonike: European Centre for Byzantine and Post-Byzantine Monuments, 2013, pp. 80–93.

Jawetz 1944a: E. Jawetz, and K. Meyer, "Studies on plague immunity in experimental animals: I. Protective and antitoxic antibodies in the serum of actively immunized animals", *Journal of Immunology* 49 (1944), pp. 1–14.

Jawetz 1944b: E. Jawetz, and K. Meyer, "Studies on plague immunity in experimental animals: II. Some factors of the immunity mechanism in bubonic plague", *Journal of Immunology* 49 (1944), pp. 15–30.

Joklik 1992: W. Joklik, "Yersinia", in W. Joklik et al. (eds), *Zinsser Microbiology*, Norwalk: Appleton & Lange, 1992, pp. 583–594.

Jorga 1896: M. Jorga, *Philippe de Mézières, 1327–1405, et la croisade au XIVe siècle* (Bibliothèque de l'École des Hautes Études 110). Paris: Baillon, 1896.

Judicial Opinion 60 1985: "Rejection of the name *Yersinia pseudotuberculosis* subsp. pestis (van Loghem) Bercovier at all 1981 and conversation of the name *Yersinia pestis* (Lehmann and Neumann) van Loghem 1944 for the plague bacillus", *International Journal of Systematic Bacteriology* 35 (1985), p. 540.

Juhas 2009: M. Juhas et al., "Genomic islands: tools of bacterial horizontal gene transfer and evolution", *Federation of European Microbiological Societies (FEMS) Microbiology Reviews* 33 (2009), pp. 376–39.

Karpozilos 1989: Α. Καρπόζηλος, "Περί αποπάτων, βόθρων και υπονόμων", in Χ. Αγγελίδη (ed.), *Η καθημερινή ζωή στο Βυζάντιο. Τομές και συνέχειες στην ελληνιστική και ρωμαϊκή παράδοση.* Αθήνα: Ινστιτούτο Ιστορικών Ερευνών/Εθνικό Ίδρυμα Ερευνών, 1999, pp. 335–352.

Kausrud 2007: K. Kausrud et al., "Climatically driven synchrony of gerbil populations allows large-scale plague outbreaks", *Proceedings of the Royal Society*, B: *Biological Sciences* 274 (2007), pp. 1963–1969.

Keeling 2000: M. Keeling, and C. Gilligan, "Bubonic plague: a metapopulation model of a zoonosis", *Proceedings of the Royal Society*, B: *Biological Sciences* 267 (2000), pp. 2219–2230.

Keeling 2008: P. Keeling, and J. Palmer, "Horizontal gene transfer in eukaryotic evolution", *Nature Reviews Genetics* 9 (8) (2008), pp. 605–618.

Keller 2018: M. Keller et al., "Ancient *Yersinia pestis* genomes from across Western Europe reveal early diversification during the First Pandemic (541–750)", *Proceedings of the National academy of Sciences of the United States of America* 116 (25) (June 18, 2019), pp. 12363–12372.

Kelly 2005: J. Kelly, *The Great Mortality.* New York, NY: Harper Collins Publishers, 2005.

Kerchen 2004: E. Kerchen et al., "The plague virulence protein YopM targets the innate immune response by causing a global depletion of NK cells" *Infection and Immunity* 72 (2004), pp. 4589–4602.

Kermack 1927: W. Kermack, and A. McKendrick, "A contribution to the mathematical theory of epidemics", *Proceedings of the Royal Society*, B: *Biological Sciences* 115 (1927), pp. 700–721.

Khan 2004: I. Khan, "Plague: the dreadful visitation occupying the human mind for centuries", *Transactions of the Royal Society of Tropical Medicine and Hygiene* 98 (2004), pp. 270–277.

Kislinger 1999: E. Kislinger, and D. Stathakopoulos, "Pest und Perserkriege bei Prokop. Chronologische Überlegungen zum Geschenen 540–545", *Byzantion* 69 (1999), pp. 76–98.

Koder 1984: J. Koder, *Der Lebensraum der Byzantiner: Historisch-geographischer Abriß ihres mittelalterlichen Staates im östlichen Mittelmeerraum.* Graz, Wien, Köln: Styria, 1984.

Konkel 2000: M. Konkel, and K. Tilly, "Temperature-regulated expression of bacterial virulence genes", *Microbes and Infections* 2 (2) (2000), pp. 157–166.

Konstantinidou 2009: K. Konstantinidou et al., "Venetian rule and control of plague epidemics on the Ionian Islands during 17th and 18th centuries", *Emerging Infectious Diseases* 15 (1) (2009), pp. 39–43.

Kordosis 1996: M. Kordosis, *Ιστορικογεωγραφικά Πρωτοβυζαντινών και εν γένει Παλαιοχριστανικών χρόνων.* Αθήνα: Καραβία, 1996.

Kordosis 2003: M. Kordosis, *Τα Βυζαντινά Γιάννενα. Κάστρο (Πόλη)-Ξώκαστρο-Κοινωνία- Διοίκηση-Οικονομία.* Αθήνα: Καραβία, 2003.

Korobeinikov 2008: D. Korobeinikov, "Raiders and neighbours: the Turks (1040–1304)", in J. Shepard (ed.), *The Cambridge History of the Byzantine Empire.* Cambridge: Cambridge University Press, 2008, pp. 703–707.

Korslund 2006: L. Korslund, and H. Steen, "Small rodent winter survival: snow conditions limit access to food resources", *Journal of Animal Ecology* 75 (2006), pp. 156–166.

Kostis 1995: K. Kostis, *Στον καιρό της πανώλης. Εικόνες από τις κοινωνίες της ελληνικής χερσονήσου 14ος:19ος αιώνας.* Ηράκλειο: Πανεπιστημιακές Εκδόσεις Κρήτης, 1995.

Koukoules 1951: Φ. Κουκουλές, *Βυζαντινών Βίος καὶ Πολιτισμός* (Collection de l'Institut français d'Athènes 73). Athènes: Institut français d'Athènes, 1951.

Kovats 2001: R. Kovats et al., "Early effects of climate change: do they include changes in vector-borne disease?", *Proceedings of the Royal Society, B: Biological Sciences* 356 (1411) (2001), pp. 1057–1068.

Krasnov 1999: B. Krasnov et al., "Additional records of fleas (Siphonaptera) on wild rodents in the southern part of Israel", *Israel Journal of Zoology* 45 (1999), pp. 333–340.

Krasnov 2001a: B. Krasnov et al., "Development rates of two Xenopsylla flea species in relation to air temperature and humidity", *Medical and Veterinary Entomology* 15 (2001), pp. 249–258.

Krasnov 2001b: B. Krasnov et al., "Effect of air temperature and humidity on the survival of pre-imaginal stages of two flea species (Siphonaptera: Pulicidae)", *Journal of Medical Entomology* 38 (2001), pp. 629–637.

Krasnov 2002: B. Krasnov et al., "Annual cycles of four flea species in the central Negev desert", *Medical and Veterinary Entomology* 16 (2002), pp. 266–276.

Krishnaswami 1972: A. Krishnaswami et al., "Investigations on plague foci in Mahasu District of Himachal Pradesh", *Indian Journal of Medical Research* 60 (8) (1972), pp. 1126–1131.

Kulikowski 2007: M. Kulikowski, "Plague in Spanish Late Antiquity", in L. Little (ed.), *Plague and the End of Antiquity. The Pandemic 541–750.* Cambridge and New York, NY: Cambridge University Press and The American Academy in Rome, 2007, pp. 150–170.

Kutyrev 2018: V. Kutyrev et al., "Phylogeny and Classification of *Yersinia pestis* through the Lens of Strains from the Plague Foci of Commonwealth of Independent States". *Frontiers in Microbiology* 9 (2018), p. 1106.

La 2008: V. La et al., "Dental Pulp as a Tool for the Retrospective Diagnosis of Infectious Diseases", in D. Raoult, and M. Drancourt (eds), *Paleomicrobiology-Past Human Infections.* Berlin and Heidelberg: Springer, 2008, pp. 175–179.

Lahodny 2013: G. Lahodny, and L. Allen, "Probability of a disease outbreak in stochastic multipatch epidemic models", *Bulletin of Mathematical Biology* 75 (7) (2013), pp. 1157–1180.

Laiou 2007: A. Laiou, and C. Morrisson, *The Byzantine Economy.* Cambridge: Cambridge University Press, 2007.

Laiou 2008: A. Laiou, "The Palaiologoi and the world around them (1261–1400)", in J. Shepard (ed.), *The Cambridge History of the Byzantine Empire*. Cambridge: Cambridge University Press, 2008, pp. 803–833.

Laird 2000: D. Laird et al., "50 million years of chordate evolution: Seeking the origins of adaptive immunity", *Proceedings of the National Academy of Science of the United States of America* 97 (13) (2000), pp. 6924–6926.

Lamb 1977: H. Lamb, *Climate: Present, Past, and Future*, volume 2: *Climatic history and the future*. London: Routledge, 1977.

Lampros 1924: Σπ. Λάμπρος, *Παλαιολόγεια καὶ Πελοποννησιακά*. Tome 2. Ἀθῆναι: Ἐπιτροπὴ ἐκδόσεως τῶν καταλοίπων Σπυρίδωνος Λάμπρου, 1924.

Lampsidis 1958: Ο. Λαμψίδης, "Μιχαὴλ Παναρέτου περὶ τῶν Κομνηνῶν", *Ἀρχεῖον Πόντου* 22 (1958), pp. 67–68.

Langer 1975: L. Langer, "The Black Death in Russia", *Russian History* 2 (1975), pp. 53–57.

Last 2001: J. Last, *Dictionary of Epidemiology*. New York, NY: Oxford University Press, 2001.

Lathem 2005: W. Lathem et al., "Progression of primary pneumonic plague: a mouse model of infection, pathology, and bacterial transcriptional activity", *Proceedings of the National Academy of Science of the United States of America* 102 (2005), pp. 17786–17791.

Lathem 2007: W. Lathem et al., "A Plasminogen-Activating Protease Specifically Controls the Development of Primary Pneumonic Plague", *Science* 315 (5811) (2007), pp. 50–513.

Le Goff 1969: J. Le Goff, and J.-N. Biraben, "La peste dans le Haut Moyen Age", *Annales. Économies, Sociétés, Civilisations* 24 (6) (1969), pp. 1484–1510.

Lemerle 1979: P. Lemerle, *Les plus anciens recueils des Miracles de Saint Démétrius et la pénétration des Slaves dans les Balkans*. I. Le texte (Le monde byzantin). Paris: Editions du Centre National de la Recherche Scientifique, 1979.

Levi 1997: M. Levi, *Logical model of plague interepizootic period*. Moscow: Materials of the VII Congress of the all-Russian Society of Epidemiologists, Microbiologists and Parasitologists, 1997, pp.79–81.

Li 2009: Y. Li et al., "Genotyping and phylogenetic analysis of *Yersinia pestis* by MLVA: insights into the worldwide expansion of Central Asia plague foci", *PLoS One* 4 (6) (2009), e6000.

Liebeschuetz 1971: J. Liebeschuetz, *Antioch-City and Imperial Administration in the Late Roman Empire*. London: Oxford University Press, 1971.

Lien-Teh 1922: W. Lien-Teh, "Plague in the Orient with special references to the Manchurian outbreaks", *Journal of Hygiene* 26 (1922), pp. 62–76.

Lien-Teh 1926: W. Lien-Teh, *A Treatise on Pneumonic Plague*. Geneva: League of Nations Health Organisation, 1926.

Lillard 1997: J. Lillard et al., "Sequence and genetic analysis of the hemin storage (hms) system of *Yersinia pestis*", *Gene* 193 (1997), pp. 13–21.

Liu 2000: Y. Liu, and J. Tan, *The Atlas of Plague and its Environment in the People's Republic of China*. Beijing: Science Press, 2000.

Llamas 2017: B. Llamas et al., "From the field to the laboratory: Controlling DNA contamination in human ancient DNA research in the high-throughput sequencing era", *STAR* 3 (1) (2017), pp. 1–14.

Loenertz 1956: R. J. Loenertz, *Démétrius Cydonès, Correspondance*, vol. 1 (Studi e Testi 186). Città del Vaticano: Biblioteca apostolica vaticana, 1956.

Lomaradski 1995: I. Lomaradski et al., "The problem of natural foci of plague: the search for ways for its resolution", *Meditsiskaya Parazitologia i Parazitarnie Bolezni (Moskva)* 4 (1995), pp. 3–9.

Lounghis 1998: T. Lounghis, "Χρονολόγηση των σταθμών εξέλιξης των μικρασιατικών θεμάτων", in V. Vlyssidou, E. Kountoura-Galake, S. Lampakes, T. Lounghis, and A. Savvides (eds), *Η Μικρά*

Ασία των Θεμάτων. Αθήνα: Εθνικό Ίδρυμα Ερευνών, Ινστιτούτο Βυζαντινών Ερευνών, 1998, pp. 37–50.

Louth 2008: A. Louth, "Byzantium Transforming (600–700)", in J. Shepard (ed.), *The Cambridge History of the Byzantine Empire.* Cambridge: Cambridge University, 2008, pp. 221–248.

Lyberopoulos 1999: Β. Λυμπερόπουλος, *Ο Βυζαντινός Πόντος: Η Αυτοκρατορία της Τραπεζούντας (1204–1461)* (Μελέτες Βυζαντινής και Μεταβυζαντινής Ελληνικής Ιστορίας 11). Αθήνα: Χαρίσης, 1999.

Macchiavello 1949: A. Macchiavello, *Nomeclature of Reservoirs and Vectors of Plague.* Geneva: World Health Organization, 1949.

Maddicott 2007: J. Maddicott, "Plague in Senventh-century England", in L. Little (ed.), *Plague and the End of Antiquity. The Pandemic 541–750.* Cambridge and NewYork, NY: Cambridge University Press, 2007, pp. 171–214.

Magdalino 1996: P. Magdalino, *Constantinople médiévale* (Travaux et mémoires du Centre d'histoire et civilization de Byzance, Collège de France, Monographies 9). Paris: De Boccard, 1996.

Maia 1952: J. Maia, "Some mathematical developments on the epidemic theory formulated by Reed and Frost", *Human Biology* 24 (1952), pp. 167–200.

Maksimovic 1994: L. Maksimovic, *Η εθνογένεση των Σέρβων στον Μεσαίωνα.* Αθήνα: Ίδρυμα Γουλανδρή-Χορν, 1994.

Malek 2017: M. Malek et al., "*Yersinia pestis* halotolerance illuminates plague reservoirs", *Scientific Reports* 7 (2017), 40022.

Mango 1980: C. Mango, *Byzantium: The Empire of New Rome.* London: Widenfeld & Nicolson, 1980.

Mango 1985: C. Mango, *Le développement urbain de Constaninople* (Travaux et mémoires du Centre de recherche d'histoire et civilisation de Byzance, Collège de France, Monographies 2). Paris: De Boccard, 1985.

Mann 1979: J. Mann et al., "Endemic Human Plague in New Mexico: Risk factors associated with infections", *Journal of Infectious Diseases* 140 (3) (1979), pp. 397–401.

Marfat 1998: B. Marfat, and J. Perret, "History of the concept of quarantine", *Médecine Tropicale* 58 (1998), pp. 14–20.

Marshall 1972a: J. Marshall et al., "The role of domestic animals in the epidemiology of plague: III. Experimental infection of wine", *Journal of Infectious Diseases* 125 (5) (1972), pp. 556–559.

Marshall 1972b: J. Marshall, D. Quy, and F. Gibson, "Asymptomatic pharyngeal plague infection in Vietnam", *American Journal of Tropical Medicine and Hygiene* 16 (2) (1972), pp. 175–177.

Mathur 2010: S. Mathur, *Statistical Bionformatics with R.* San Diego, CA, and London: Elsevier-Academic Press, 2010.

Martinevsky 1973: I. Martinevsky, "Materials for typing natural foci of plague according to the genetic properties of plague bacillus strains", *Journal of Hygiene, Epidemiology, Microbiology, and Immunology*17 (3) (1973), pp. 272–278.

May 2001: C. May, *Basic Laboratory Protocols for the Presumptive Identification of* Yersinia pestis. Atlanta, GA: Centers for Disease Control and Prevention CDC, 2001, pp. 1–19.

Mayerson 1996: P. Mayerson, "Egeria and Peter the Deacon on the Site of Clysma (Suez)", *Journal of the American Research Center in Egypt* 33 (1996), pp. 61–64.

McCaa 1995: R. McCaa, "Spanish and Nahuatl view on smallpox and demographic catastrophe in the conquest of Mexico", *Journal of Interdisciplinary History* 25 (3) (1985), pp. 397–431.

McCormick 2003: M. McCormick, "Rats, Communications and plague: Toward an Ecological History", *Journal of Interdisciplinary History* 34 (1) (2003), pp. 1–25.

McCormick 2007: M. McCormick, "Toward a Molecular History of the Justinianic Pandemic" in L. Little (ed.), *Plague and the End of Antiquity: Pandemic of 541–750*. Cambridge and New York, NY: Cambridge University Press and The American Academy in Rome, 2007, pp. 290–301.

McMichael 2003: A. McMichael et et al. (eds), *Climate change and human health: risks and responses*. Geneva: World Health Organization, 2003.

McNeill 1976: W. McNeill, *Plagues and Peoples*. New York, NY: Anchor Press, 1976.

Mears 2002: S. Mears et al., "Host location, survival and fecundity of the Oriental rat flea *Xenopsylla cheopis* (Siphonaptera Pulicidae) in relation to black rat *Rattus rattus* (Rodentia: Muridae) host age and sex", *Bulletin of Entomological Research* 92 (5) (1976), pp. 375–384.

Metcalf 2015: C. Metcalf et al., "Understanding Herd Immunity", *Trends inImmunology* 36 (12) (2015), pp. 753–755.

Meyer 1950: K. Meyer, "Immunity in plague: a critical consideration of some recent studies", *Journal of Immunology* 64 (1950), pp. 139–163.

Michaud 1852: J. Michaud, *History of the Crusades*. London: G. Routledge, 1852.

Mikhailova 2001: M. Mikhailova, "Hydrological Regime of the Nile Delta and Dynamics of its Coastline", *Water Resources* 28 (2001), pp. 526–539.

Milani 1977: C. Milani, *Itinerarium Antonini Placentini. Un viaggio in terra Santa del 560–570 d.C.* Milano: Vita e Pensiero, 1977.

Miller 1908: W. Miller, *The Latins in the Levant: A History of Frankish Greece (1204–1566)*. London: John Murray, 1908.

Miller 1976: T. Miller, "The plague in John VI Cantacuzenus and Thucydides", *Greek, Roman and Byzantine Studies* 17 (1976), pp. 385–395.

Miquel 1977: A. Miquel, *L'Islam et sa civilization VIIe-XXe siècle*. Paris: Armand Colin, 1977.

Molina 2010: P. Molina et al., "Focus on: Alcohol and the immune system", *Alcohol Research and Health* 33 (2010), pp. 97–108.

Mollaret 1963: H. Mollaret, "Experimental preservation of plague in soil", *Bulletin de la Société de Pathologie Exotique* 56 (1963), pp. 1168–1182.

Mollaret 1969: H. Mollaret, "The Ark of the Covenant and the disease of Philistines: Dysentery, plague or parasitic disease?", *La Presse Médicale* 77 (55) (1969), pp. 2111–2114.

Möller 2018: S. Möller, L. Plessi, and T. Stadler, "Impact of the tree prior on estimating clock rates during epidemic outbreaks", *Proceedings of the National Academy of Science of the United States of America* 115 (16) (2018), pp. 4200–4205.

Montanari 2015: F. Montanari, *The Brill Dictionary of Ancient Greek*. M. Goh, and C. Schroeder (eds). Leiden: Brill, 2015.

Moorhead 2008: J. Moorhead, "Western Approaches (500–600)", in J. Shepard (ed.), *The Cambridge History of the Byzantine Empire*. Cambridge: Cambridge University, 2008, pp. 203–212.

Mordechai 2019a: L. Mordechai, and M. Eisenberg, "Rejecting Catastrophe: The Case of the Justinianic Plague", *Past & Present* 244 (1) (2019), pp. 3–50.

Mordechai 2019b: L. Mordechai et al., "The Justinian Plague: An inconsequential pandemic? ", *Proceedings of the National Academy of Sciences of the United States of America* 116 (51) (2019), pp. 25546–25554.

Moreira-Texeira 2017: L. Moreira-Texeira et al., "Cell-derived IL-10 impairs host resistance to *Mycobacterium tuberculosis* infection", *Journal of Immunology* 199 (2) (2017), pp. 613–623.

Morelli 2010: G. Morelli et al., "Phylogenetic diversity and historical patterns of pandemic spread of *Yersinia pestis*", *Nature Genetics* 42 (12) (2010), pp. 1140–1143.

Morens 1992: D. Morens, and R. Littman, "Epidemiology of the Plague of Athens", *Transactions of American Philosophical Society* 122 (1992), pp. 271–304.

Morens 1994: D. Morens, and R. Littman, "Thucydides Syndrome reconsidered: new thoughts on the the Plague of Athens", *American Journal of Epidemiology* 140 (7) (1994), pp. 621–630.

Morens 2009: D. Morens, and J. Taubenberger, "Understanding influenza backward", *Journal of the American Medical Association* 302 (6) (2009), pp. 679–680.

Morony 2007: M. Morony, "For whom does the writer write? The first bubonic plague pandemic according to Syriac sources", in L. Little (ed.), *Plague and the End of Antiquity. The Pandemic 541–750*. Cambridge and New York, NY: Cambridge University Press and The American Academy in Rome, 2007, pp. 59–86.

Motin 2002: V. Motin et al., "Genetic variability of *Yersinia pestis* isolates as predicted by PCR-based IS*100* genotyping and analysis of structural genes encoding glycerol-3-phosphate dehygrogenase (*glpD*)", *Journal of Bacteriology* 184 (4) (2002), pp. 1019–1027.

Murray 1954: A. Murray, *Homer, The Iliad*. Cambridge, MA: Harvard University Press, 1954.

Nachtergael 1988: G. Nachtergael, "Un aspect de l'environnement en Égypte gréco-romaine: les dangers de la circulation", *Ludus Magistralis* 2 (1988), pp. 19–54.

Namouchi 2018: A. Namouchi et al., "Integrative approach using *Yersinia pestis* genomes to revisit the historical landscape of plague during the Medieval Period", *Proceedings of the National Academy of Science of the United States of America* 115 (50) (2018), E11790–E11797.

Nans 2015: A. Nans et al., "Structure of a bacterial type III secretion system in contact with a host membrane *in situ*", *Nature Communications* 6 (2015), 10114.

Nedialkov 1997: Y. Nedialkov, V. Motin, and R. Brubaker, "Resistance to lipopolysaccharide mediated by the *Yersinia pestis* V antigen-polyhistidine fusion peptide: amplification of interleukin-10", *Infection and Immunity* 65 (1997), pp. 1196–1203.

Nguyen 2018: V. Nguyen, C. Parra-Rojas, and E. Hernadez-Vargas, "The 2017 plague outbreak in Madagascar: data descriptions and epidemic modeling", *Epidemics* 25 (2018), pp. 20–25.

Nicol 1991: D. Nicol, *A Biographical Dictionary of the Byzantine Empire*. London: Seaby, 1991.

Nishiura 2006: H. Nishiura, "Epidemiology of a primary pneumonic plague in Kantoshu, Manchuria, from 1910 to 1911: statistical analysis of individual records collected by the Japanese Empire", *International Journal of Epidemiology* 35 (4) (2006), pp. 1059–1065.

Nissapatorn 2006: V. Nissapatorn et al., "Tuberculosis: a resurgent disease in immunosuppressed patients", *Southeast Asian Journal of Tropical Medicine and Public Health* 37 (Suppl 3) (2006), pp. 153–60.

Nistazopoulou-Pelekidi 1978: Μ. Νυσταζοπούλου-Πελεκίδη, " Ἡ ἀνόρθωση 802–945", in Γ. Χριστόπουλος, Ἰ. Μπαστιᾶς (eds), *Ἱστορία τοῦ Ἑλληνικοῦ Ἔθνους. Τόμος Ζ': Βυζαντινὸς Πολιτισμός. Πρωτοβυζαντινοὶ Χρόνοι*. Ἀθῆνα: Ἐκδοτικὴ Ἀθηνῶν, 1978, pp. 46–97.

Nogales 2018: A. Nogales et al., "Functional evolution of the 2009 Pandemic H1N1 Influenza Virus NS1 and PA in humans", *Journal of Virology* 92 (19) (2018), e01206–18.

Noiret 1892: H. Noiret, *Documents inédits pour servir à l'histoire de la domination vénitienne en Crète de 1380 à 1485. Tirés des archives de Venise* (Bibliothèque des Écoles françaises d'Athènes et de Rome 61) Paris: Thorin & fils, 1892.

Norris 1977: J. Norris, "East or West? The geographic origin of the Black Death", *Bulletin of the History of Medicine* 51 (1) (1977), pp. 1–24.

Novick 2016: R. Novick, and G. Ram, "The floating (pathogenicity) island: a genomic dessert", *Trends in Genetics* 32 (2) (2016), pp. 114–126.

Obituary 1931: "Baron Shibasaburo Kitasato", *British Medical Journal* 1 (1931), pp. 1141–1142.

Ochman 1987: H. Ochman and A. Wilson, "Evolution in bacteria: evidence for a universal substitution rate in cellular genome", *Journal of Molecular Evolution* 26 (1987), pp. 74–86.

Ochman 2000: H. Ochman, J. Lawrence, and E. Groisman, "Lateral gene transfer and the nature of bacterial innovation", *Nature* 405 (6784) (2000), pp. 299–304.

Official Publications Received 1933: "Alcoholism in Medieval England", *Nature* 132 (130) (1933), p. 130.

Olea 2005: R. Olea, and G. Christakos, "Duration of Urban Mortality for the 14th-century Black Death Epidemic", *Human Biology* 77 (3) (2005), pp. 291–303.

Orlanski 1975: J. Orlanski, "A rational subdivision of scales for atmospheric processes", *Bulletin of the American Mathematical Society* 56 (1975), pp. 529–530.

Ostrogorsky 1968: G. Ostrogorsky, *History of the Byzantine State.* Oxford: Blackwell, 1968.

Paidousis 1939: Μ. Παϊδούσης, " Ἐπιδημήσαντα καὶ ἐπιδημοῦντα νοσήματα εἰς Χίον", *Περιοδικὸν τοῦ ἐν Χίῳ Συλλόγου Ἀργέντη* 8 (1939), p. 46.

Palmer 1965: W. Palmer, *Meteorological Drought Research Paper No.45.* Washington D.C.: US Weather Bureau, 1965.

Palmer 1994: D. Palmer, "Plague and other *Yersinia* infections", in K. Isselbacher et et al. (eds), *Harrison's Principles of Internal Medicine.* New York: Mc Graw-Hill, 1994, pp. 957–983.

Papagrigorakis 2006: M. Papagrigorakis et al., "DNA examination of ancient dental pulp incriminates typhoid fever as a probable cause of the Plague of Athens", *International Journal of Infectious Diseases* 10 (3) (2006), pp. 206–214.

Paradis 2005: S. Paradis et al., "Phylogeny of the Enterobacteriaceae on genes encoding elongation factor Tu and F-ATP-beta unit", *Internationa Journal of Systematic and Evolutionary Microbiology* 55 (5) (2005), pp. 2013–2025.

Park 2007: S. Park et al., "Statistical analysis of the dynamics of antibody loss to a disease-causing agent: plague in natural population of great gerbils as an example", *Journal of the Royal Society of Interface* 4 (12) (2007), pp. 57–64.

Parker 2002: S. Parker, "The Roman Frontier in Jordan: an overview", in P. Freeman, J. Bennett, Z. T. Fiema, B. Hoffmann (eds.), *Limes XVI. Proceedings of the XVIIIth International Congress of Roman Frontier Studies*, volume 1 (BAR International Series 1084). Oxford: British Archaeological Reports, 2002, pp. 77–84.

Parkhill 2001: J. Parkhill et al., "Genome sequence of *Yersinia pestis*, the causative agent of plague", *Nature* 413 (6855) (2001), pp. 523–527.

Parmenter 1999: R. Parmenter et al., "Incidence of plague associated with increased winter-spring precipitation in New Mexico", *American Journal of Tropical Medicine and Hygiene* 61 (1999), pp. 814–821.

Patoura-Spanou 2008: Σ. Πατούρα-Σπανού, *Η μεθόριος του Δούναβη και ο κόσμος της στην εποχή της μετανάστευσης των λαών (4ος:7ος αι.).* Αθήνα: Εθνικό Ἰδρυμα Ερευνών, Ινστιτούτο Ιστορικών Ερευνών, 2008.

Patz 1996: J. Patz et al., "Global climate change and emerging infectious diseases", *Journal of the American Medical Association* 275 (1996), pp. 217–223.

Pechous 2016: R. Pechous et al., "Pneumonic Plague: the darker side of *Yersinia pestis*", *Trends in Microbiology* 24 (3) (2016), pp. 190–197.

Perry 1997: R. Perry, and J. Fetherston, "*Yersinia pestis* –etiologic agent of plague", *Clinical Microbiology Review* 10 (1) (1997), pp. 35–66.

Perry 1999: R. Perry et al., "The hmu locus of *Yersinia pestis* and analysis of hmu mutants for hemin and hemoprotein utilization", *Infection and Immunity* 67 (1999), pp. 3879–3892.

Pham 2009: H. Pham et al., "Correlates of environmental factors and human plague: An ecological study in Vietnam", *International Journal of Epidemiology* 38 (2009), pp. 1634–1641.

Philip 1921: A. Philip, *The Calendars: its history, structure and improvement.* Cambridge: Cambridge University Press, 1921.

Pike 1976: R. Pike, "Laboratory-associated infections: Summary and analysis of 3921 cases", *Health Laboratory Science* 13 (1976), pp. 105–114.

Plano 2013: G. Plano, and K. Schesser, "The *Yersinia pestis* type III secretion system: expansion, assembly and role in the evasion of host defenses", *Immunologic Research* 2013 (1–3), 10.1007.

Poland 1973: J. Poland, A. Barnes, and J. Herman, "Human bubonic plague from exposure to a naturally infected wild carnivore", *American Journal of Epidemiology* 97 (5) (1973), pp. 332–337.

Poland 1994: J. Poland, J. Quan, and A. Barnes, "Plague", in G. W. Beran (ed.), *Handbook of zoonoses. Section A. Bacterial, Rickettsial, and Mycotic Diseases*. Boca Raton, FL: CRC Press, 1994, pp. 93–112.

Poland 1999: J. Poland, and D. Dennis, "Diagnosis and Clinical Manifestations", in D. Dennis (ed.), *Plague Manual: Epidemiology, Distribution, Surveillance and Control*. Geneva: World Health Organization. Communicable Disease Surveillance and Response (WHO/CDS/CSR/EDC/99.2), 1999, pp. 43–53.

Pollitzer 1954: R. Pollitzer, *Plague*. Geneva: World Health Organization, 1954.

Pouillot 2008: F. Pouillot, C. Fayolle, and E. Carniel, "Characterization of chromosomal regions conserved in *Yersinia pseudotuberculosis* and lost by *Yersinia pestis*", *Infection and Immunity* 76 (10) (2008), pp. 4592–4599.

Pourcel 2004: C. Pourcel et al., "Tandem repeats analysis for the high-resolution phylogenetic analysis of *Yersinia pestis*", *BMC Microbiology* 4 (2004), 22.

Prakash 2002: O. Prakash et al., "Hepatitis C virus (HCV) and human immunodeficiency virus type 1 (HIV-1) infections in alcoholics", *Frontiers in Bioscience* 7 (2002), e286–300.

Preiser-Kapeller 2013: J. Preiser-Kapeller, "Mapping maritime networks in Byzantium: Aims and prospects of the project Ports and landing places at the Balkan coasts of the Byzantine Empire", in F. Karagiani (ed.). *Proceedings of International Symposium "Medieval Ports in North Aegean and the Black Sea: Links to the Maritime Routes of the East" in the frame of the project "OLKAS: From Aegean to the Black Sea Medieval ports in the Maritime Routes of the East"*. Thessalonike: European Centre for Byzantine and Post-Byzantine Monuments, 2013, pp. 467–492.

Preiser-Kapeller 2014: J. Preiser-Kapeller, and F. Daim (eds), *Harbours and Maritime Networks as Complex Adaptive Systems* (RGZM–Tagungen 23; Interdisziplinäre Forschungen zu den Häfen von der Römischen Kaiserzeit bis zum Mittelalter in Europa, Band 2). Mainz: Römisch-Germanisches Zentralmuseum, 2014.

Price 1995: S. Price, M. Freeman, and K. Yeh, "Transcriptional analysis of the *Yersinia pestis* ph6 antigen gene", *Journal of Bacteriology* 117 (1995), pp. 5997–6000.

Qi 2016: Z. Qi et al., "Taxonomy of *Yersinia pestis*", in R. Yang, and A. Anisimov (eds), *Yersinia pestis: Retrospective and Perspective* (*Advances in Experimental Medicine and Biology* 918). Dordrecht: Springer, 2016, pp. 35–78.

Radnedge 2002: L. Radnedge et al., "Genome lasticity in *Yersinia pestis*", *Microbiology* 148 (2002), pp. 1687–1698.

Randall 2002: J. Randall, and K. Rogovin, "Variation in and meaning of alarm calls in a social desert rodent *Rhombomys opimus*", *Ethology* 108 (2002), pp. 513–527.

Raoult 2000: D. Raoult et al., "Molecular identification by "suicide PCR" of *Yersinia pestis* as the agent of medieval black death", *Proceedings of the National Academy of Science of the United States of America* 97 (23) (2000), pp. 12800–12803.

Raoult 2013: D. Raoult et al., "Plague: History and contemporary analysis", *Journal of Infection* 66 (1) (2013), pp. 18–26.

Rascovan 2019: N. Rascovan et al., "Emergence and Spread of Basal Lineages of *Yersinia pestis* during the Neolithic Decline", *Cell* 176 (1–2) (2019), pp. 295–305.

Rasmussen 2015: S. Rasmussen et al., "Early divergent strains of *Yersinia pestis* in Eurasia 5,000 years ago", *Cell* 163 (3) (2015), pp. 571–582.

Ratledge 2000: C. Ratledge, and L. Dover, "Iron metabolism in pathogenic bacteria", *Annual Review of Microbiology* 54 (2000), pp. 881–941.

Reuter 2014: S. Reuter et al., "Parallel independent evolution of pathogenicity within the genus *Yersinia*", *Proceedings of the National Academy of Science of the United States of America* 111 (18) (2014), pp. 6768–6773.

Riffaat 1981: M. Riffaat, T. Morsy, and M. Abdel-Mawla, "Seasonal activity of *Rattus norvegicus* and flea index in Port Said Governorate", *Journal of the Egyptian Society of Parasitology* 11 (2) (1981), pp. 525–532.

Riley 2018: L. Riley, and R. Blanton, "Advances in Molecular Epidemiology of Infectious Diseases: definitions, approaches and scope of the field", *Microbiology Spectrum* 6 (6) (2018), pp. 1–18.

Riley 2019: L. Riley, "Differentiating Epidemic from Endemic or Sporadic Infectious Diseases occurence", *Microbiology Spectrum* 7 (4) (2019), pp. 1–16.

Roberts 2007: T. Roberts et al., "Immunosuppression during Active Tuberculosis is Characterized by Decreased Interferon-g Production and CD25 Expression with Elevated Forkhead Box P3, Transforming Growth Factor–b, and Interleukin-4 mRNA Levels", *Journal of Infectious Diseases* 195 (2007), pp. 870–878.

Rochow 1991: I. Rochow, *Byzanz im 8 Jahrhundert in der Sicht des Theophanes: quellenkritisch-historicher Kommentar zu den Jahren 715–813* (Berliner Byzantinistische Arbeiten 57). Berlin: Akademie Verlag, 1991.

Rogers 1928: L. Rogers, "The yearly variations in plague in India in relation to climate: forecasting epidemics", *Proceedings of the Royal Society*, B: *Biological Sciences* 103 (1928), pp. 42–72.

Roosen 2019: J. Roosen, and D. Curtis, "The '*light touch*' of the Black Death in the Southern Netherlands: an urban trick?", *Economic History Review* 72 (1) (2019), pp. 32–56.

Rose 2003: L. Rose. et al., "Survival of *Yersinia pestis* upon envrimental surfaces", *Applied and Environmental Microbiology* 63 (4) (2003), pp. 2166–2171.

Rosenberg 1968: J. Rosenberg, "Alexandre Yersin", *New England Journal of Medicine* 278 (5) (1968), pp. 261–263.

Rothschild 1972: M. Rothschild et al., "Jump of the Oriental rat flea *Xenopsylla cheopis* (Roths.)", *Nature* 239 (1972), pp. 45–48.

Rougé 1952: J. Rougé, " La navigation hivernale sous l'Empire Romain", *Revue des études anciennes* 54 (1952), pp. 316–325.

Ruffié 1984: J. Ruffié, and C. Sournia, *Les épidémies dans l'histoire de l'homme*. Paris: Flammarion, 1984.

Runciman 1958: S. Runciman, *The Sicilian Vespers*. Cambridge: Cambridge University Press, 1958.

Runciman 1980: S. Runciman, *Mistra, Byzantine capital of the Peloponnese*. London: Thames and Hudson, 1980.

Russell 1976: J. Russell, "The earlier medieval plague in the British Isles", *Viator* 7 (1976), pp. 65–78.

Rust 1971: J. Rust et al., "The role of domestic animals in the epidemiology of plague. II. Antibody to *Yersinia pestis* in sera of dogs and cats", *Journal of Infectious Diseases* 124 (5) (1971), pp. 527–531.

Sabbatani 2009: S. Sabbatani, and S. Fiorino, "The Antonine Plague and the decline of the Roman Empire", *Infezzioni in Medicina* 17 (4) (2009), pp. 261–275.

Sabina 2011: Y. Sabina et al., "*Yersinia enterocolitica:* Mode of Transmission, Molecular Insights of Virulence, and Pathogenesis of Infection", *Journal of Pathogens* 2011, 429069.

Sallares 2007: R. Sallares, "Ecology, evolution, and epidemiology of plague", in L. Little (ed.), *Plague and the End of Antiquity: The Pandemic of 541–750*. Cambridge and New York, NY: Cambridge University Press and The American Academy in Rome, 2007, pp. 231–289.

Salinger 2005: J. Salinger, M. Sivakumar, and R. Motha, "Reducing vulnerability of agriculture and forestry to climate climate variability and change", *Climatic Change* 70 (1–2) (2005), pp. 341–362.

Salomonsson 2016–2018: E. Salomonsson et al., *Improved methods and capability for laboratory biopreparedness*. Stockholm: Swedish Forum for Biopreparedness Diagnostics, 2016–2018.

Sarris 2007: P. Sarris, "Bubonic Plague in Byzantium: The Evidence of Non-Literary Sources", in L. Little (ed.), *Plague and the End of Antiquity: The Pandemic of 541–750*. Cambridge and New York, NY: Cambridge University Press and The American Academy in Rome, 2007, pp. 119–134.

Sathas 1880: Konstantinos Sathas, *Documents inédits relatifs à l'histoire de la Grèce*, 9 vols. Paris: Maisonneuve, 1880–1890.

Savvides 1983: Α. Σαββίδης, *Τα χρόνια της σχηματοποίησης του Βυζαντίου (284–518 μ.Χ.)*. Αθήνα: Βασιλόπουλος, 1983.

Sbeinati 2010: M. Sbeinati et al., "Timing of earthquake ruptures at the Al-Harif Roman aqueduct (Dead Sea Fault, Syria) from archaeoseismology and paleoseismology", in M. Sintubin et al. (eds), *Ancient Earthquakes*. Boulder, CO: Geological Society of America, 2010, pp. 243–245.

Schaechter 2006: M. Schaechter, J. Ingraham, and F. Neidhardt, *Microbe*. Washington D.C.: American Society for Microbiology, 2006.

Schleifstein 1939: J. Schleifstein, and M. Coleman, "An unidentified microorganism resembling *Bacterium lignieri* and *Pasteurella pseudotuberculosis* and pathogenic for man", *New York State Journal of Medicine* 39 (1939), pp. 1749–1753.

Schmid 2015: B. Schmidet al., "Climate-driven introduction of the Black Death and successive plague reintroductions into Europe", *Proceedings of the National Academy of Science of the United States of America* 112 (10) (2015), pp. 3020–3025.

Schmidt 2004: H. Schmidt, and M. Hensel, "Pathogenicity Islands in Bacterial Pathogenesis", *Clinical Microbiology Review* 17 (1) (2004), pp. 14–56.

Schneider 2009: W. Schneider, "Smallpox in Africa during Colonial Rule", *Medical History* 53 (2009), pp. 193–227.

Schönberger 1969: H. Schönberger, "The Roman Frontier in Germany: an Archaeological Survey", *Journal of Roman Studies* 59 (1969), pp. 144–197.

Schwiesow 2015: L. Schwiesow et al., "*Yersinia* Type III Secretion System Master Regulator *LcrF*", *Journal of Bacteriology* 198 (4) (2015), pp. 604–614.

Scott 2001: S. Scott, and C. Duncan, *Biology of plagues: evidence from historical populations*. Cambridge: Cambridge University Press, 2001.

Seal 1969: S. Seal, "Epidemiological studies of the plague in India: the present position", *Bulletin of the World Health Organization* 23 (1969), pp. 283–292.

Seecharran 2017: T. Seecharran et al., "Phylogeographic separation and formation of sexually discrete lineages in a global population of *Yersinia pseudotuberculosis*", *Microbial Genome* 3 (10) (2017), e000133.

Setton 1975: K. Setton, *Catalan Domination of Athens, 1311–1388*. London: Variorum, 1975.

Shaddock 2014: E. Shaddock et al., "Secondary bacterial infection in active pulmonary tuberculosis", *Southern African Journal of Infectious Diseases* 29 (1) (2014), pp. 23–26.

Shames 1995: R. Shames, and D. Adelman, "Disorders of the Immune System", in S. McPhee et al. (eds), *Pathophysiology of Disease: an introduction to Clinical Medicine*. London: Appleton & Lange, 1995, pp. 28–48.

Shapiro 2006: B. Shapirom, and T. Gilbert, "No proof that typhoid caused the Plague of Athens (a reply to Papagrigorakis et et al.)", *International Journal of Infectious Diseases* 10 (2006), pp. 334–340.

Sheppard 2010: P. Sheppard, "Dendroclimatology: extracting climate from trees", *WIREs Climatic Change* 1 (3) (2010), pp. 343–352.

Sherman 2017: I. Sherman, *The power of plagues*. Washington, D.C.: American Society for Microbiology, 2017.

Shoukry 1986: A. Shoukry et al., "Seasonal activities of two commensal rats and flea index in North Sinai Governorate", *Journal of the Egyptian Society of Parasitology* 16 (2) (1986), pp. 385–393.

Shulman 2004: H. Shulman, M. Reshef, and Z. Ben-Avraham, "The structure of the Golan Heights and its tectonic linkage to the Dead Sea Transform and the Palmyrides folding", *Israel Journal of Earth Sciences* 53 (2004), pp. 225–237.

Sidebotham 1991: S. Sidebotham, "Ports of the Red Sea and the Arabia-India trade", in V. Begley, and R. De Puma (eds), *Rome and India: The Ancient Sea Trade* (Wisconsin Studies in Classics). Madison, WI: University of Wisconsin Press, 1991, pp. 12–38.

Sidebotham 1996: S. Sidebotham, and R. Zitterkopf, "Survey of the Via Hadriana: the 1996 season", *Bulletin de l'Institut français d'archéologie orientale* 97 (1996), pp. 221–237.

Sidebotham 1997: S. Sidebotham, "Caravans across the Eastern Desert of Egypt: recent discoveries on the Berenike-Apollonopolis Magna-Koptos road", in A. Avanzini, and G. Salmeri (eds), *Profumi d'Arabia: Atti del convegno* (Saggi di storia antica). Roma: L'Erma di Bretschneider, 1997, pp. 385–393.

Sidebotham 2002: S. Sidebotham, and W. Wendrich, "Berenike: Archaeological fieldwork at a Ptolemaic-Roman port on the Red Sea coast of Egypt 1999–2000", *Sahara* 13 (2002), pp. 31–44.

Sijpesteijn 1963: P. Sijpesteijn, "Der ΠΟΤΑΜΟΣ ΤΡΑΙΑΝΟΣ", *Aegyptus* 43 (1963), pp. 70–83.

Sinakos 2003: Α. Σινάκος, *Άνθρωπος και περιβάλλον στην Πρωτοβυζαντινή εποχή (4ος-6ος αι.)*. Θεσσαλονίκη: University Studio Press, 2003.

Sing 2002a: A. Singet al., "*Yersinia* V-antigen exploits Toll-like receptor 2 and CD14 for interleukin 10-mediated immunosuppression", *Journal of Experimental Medicine* 196 (2002), pp. 1017–1024.

Sing 2002b: A. Sing et al., "*Yersinia enterocolitica* evasion of the host immune response by V antigen-induced IL-10 production of macrophages is abrogated in IL-10-deficient mice", *Journal of Immunology* 168 (2002), pp. 1315–1321.

Singleton 1989: G. Singleton, "Population dynamics of an outbreak of house mouse (*Mus domesticus*) in the mallee wheatlands of Australia—hypothesis of plague formation", *Journal of Zoology (London)* 219 (1989), pp. 495–515.

Skerman 1980: V. Skerman, V. McGowan, and P. Sneath, "Approved lists of bacterial names", *International Journal of Systematic Bacteriology* 30 (1980), pp. 225–420.

Slack 1980: P. Slack, "The disappearance of plague: an alternative view", *Economic History Review* 34 (1980), pp. 469–476.

Smiley 2008: S. Smiley, "Immune defense against pneumonic plague", *Immunological Reviews* 225 (2008), pp. 256–271.

Smith 1996: M. Smith, and N. Thanh, "Plague", in G. Cook (ed.), *Manson's Tropical Diseases*. London: Saunders Company, 1996, pp. 918–924.

Snäll 2008: T. Snäll et al., "Climate-driven spatial dynamics of plague among prairie dog colonies", *American Naturalist* 171 (2008), pp. 238–248.

Sneh 1973: A. Sneh, and T. Weissbrod, "Nile Delta: The defunct Pelusiac branch identified", *Science* 180 (4081) (1973), pp. 59–61.

Solomon 1997: T. Solomon, "Hong Kong 1894: the role of James A. Lowson in the controversial discovery of the plague bacillus", *Lancet* 350 (1997), pp. 59–62.

Song 2004: Y. Song et al., "Complete genome sequence of *Yersinia pestis* strain 91001, an isolate avirulent to humans", *DNA Research* 11 (3) (2004), pp. 179–190.

Spyrou 2016: A. Spyrou et al., "Historical *Yersinia pestis* genomes reveal the European Black Death as the sources of Ancient and Modern Plague Pandemics", *Cell Host and Microbe* 19 (2016), pp. 874–881.

Srivastava 1978: L. Srivastava et al., "Serological surveillance of plague in endemic areas of South India", *Indian Journal of Medical Research* 68 (7) (1978), pp. 12–15.

Staggs 1991: T. Staggs, and R. Perry, "Identification and cloning of a fur regulatory gene in *Yersinia pestis*", *Journal of Bacteriology* 173 (1991), pp. 417–425.

Stathakopoulos 2004: D. Stathakopoulos, *Famine and pestilence in the Late Roman and Early Byzantine Empire—A systematic survey of subsistence crises and epidemics* (Birmingham Byzantine and Ottoman Monographs 9). Aldershot and Burlington, VT: Ashgate, 2004.

Stanley 2008: J.-D. Stanley, M. Bernasconi, and T. Jorstad, "Pelusium, an Ancient Port Fortress on Egypt's Nile Delta Coast: its evolving environmental setting from foundation to demise", *Journal of Coastal Research* 24 (2) (2008), pp. 451–462.

Stapp 2004: P. Stapp, M. Antolin, and M. Ball, "Patterns of extinction in prairie dog metapopulations: plague outbreaks follow El Niño events", *Frontiers in Ecology and Environment* 2 (2004), pp. 235–240.

Stenseth 2006: N. Stenseth et al., "Plague dynamics are driven by climate variation", *Proceedings of the National Academy of Science of the United States of America* 103 (35) (2006), pp. 13110–13115.

Stenton 1971: F. Stenton, *Anglo-Saxon England*. Oxford: Oxford University Press, 1971.

Stephenson 2008: P. Stephenson, "Balkan borderlands (1018–1204)", in J. Shepard (ed.), *The Cambridge History of the Byzantine Empire*. Cambridge: Cambridge University Press, 2008, pp. 678–690.

Strand 1977: M. Strand, "Pathogens of *Siphonaptera* (Fleas)", in D. Roberts, and M. Strand (eds), *Pathogens of Medically Important Arthropods*. Geneva: World Health Organization, 1977, pp. 279–359.

Stratos 1978: Α. Στράτος, "Ἡ "τυρρανίς" τοῦ Φωκᾶ", in Γ. Χριστόπουλος, Ἰ. Μπαστιᾶς (eds), *Ἱστορία τοῦ Ἑλληνικοῦ Ἔθνους. Τόμος Ζ': Βυζαντινὸς Πολιτισμός. Πρωτοβυζαντινοὶ Χρόνοι.* Ἀθῆνα: Ἐκδοτικὴ Ἀθηνῶν, 1978, pp. 222–243.

Svoronos 1978: Ν. Σβορῶνος, "Οἰκονομία-Κοινωνία", in Γ. Χριστόπουλος, Ἰ. Μπαστιᾶς (eds), *Ἱστορία τοῦ Ἑλληνικοῦ Ἔθνους. Τόμος Ζ': Βυζαντινὸς Πολιτισμός. Πρωτοβυζαντινοὶ Χρόνοι.* Ἀθῆνα: Ἐκδοτικὴ Ἀθηνῶν, 1978, pp. 278–305.

Sygkellou 2008: Ε. Συγκέλλου, *Ο Πόλεμος στο Δυτικό Ελλαδικό Χώρο κατά τον Ύστερο Μεσαίωνα (13ος–15ος αιώνας).* Αθήνα: Ινστιτούτο Ιστορικών Ερευνών/Εθνικό Ίδρυμα Ερευνών, 2008.

Szabo 2009: G. Szabo, and P. Mandrekar, "A recent perspective on alcohol, immunity, and host defense", *Alcoholism, Clinical and Experimental Research* 33 (2009), pp. 220–232.

Taubenberger 2009: J. Taubenberger, and D. Morens, "Pandemic influenza—including a risk assessment of H5N1", *Revue scientifique et technique* 28 (1) (2009), pp. 187–202.

Taylor 2014: M. Taylor, *Viruses and Man: A History of Interactions*. New York, NY: Springer, 2014.

Taylor 2017: R. Taylor, *The Amazing Language of Medicine: Understanding Medical Terms and their Backstories*. New York, NY: Springer, 2017.

Tegel 2010: W. Tegel, J. Vanmoerkerke, and U. Büntgen, "Updating historical tree-ring records from climate reconstruction", *Quaternary Science Reviews* 29 (2010), pp. 1957–1959.

Telelis 2004: Ι. Τελέλης, *Μετεωρολογικά φαινόμενα και κλίμα στο Βυζάντιο.* Αθήνα: Ακαδημία Αθηνῶν, Κέντρον Ερεύνης της Ελληνικής και Λατινικής Γραμματείας, 2004.

Thiriet 1958: F. Thiriet, *Régestes des délibérations du Sénat de Venise concernant la Roumanie (1329–1399)*, volume 1 (École Pratique des Hautes Études, Sorbonne, VIe Section. Documents et recherches sur l'économie des pays byzantins, islamiques et slaves et leurs relations commerciales au Moyen Age 1). Paris: Mouton, 1958.

Thiriet 1959: F. Thiriet, *Régestes des délibérations des assemblées vénitiennes concernant la Roumanie (1400–1430)*, volume 2 (École Pratique des Hautes Études, Sorbonne, VIe Section. Documents et recherches sur l'économie des pays byzantins, islamiques et slaves et leurs relations commerciales au Moyen Age 2). Paris: Mouton, 1959.

Thiriet 1961: F. Thiriet, *Régestes des délibérations du Senat de Venise concernant la Roumanie (1431–1463)*, volume 3 (École Pratique des Hautes Études, Sorbonne, VIe Section. Documents et recherches sur l'économie des pays byzantins, islamiques et slaves et leurs relations commerciales au Moyen Age 4). Paris: Mouton, 1961.

Thiriet 1966: F. Thiriet, *Délibérations des assemblées vénitiennes concernant la Roumanie (1160–1363)*, volume 1 (École Pratique des Hautes Études, Sorbonne, VIe Section. Documents et recherches sur l'économie des pays byzantins, islamiques et slaves et leurs relations commerciales au Moyen âge 8). Paris: Mouton, 1966.

Titball 2001: R. Titball, and E. Williamson, "Vaccination against bubonic and pneumonic plague", *Vaccine* 19 (2001), pp. 4175–4184.

Tononi 1884: A. Tononi, "La peste dell'anno 1348", *Giornale Ligustico di Archologia, Storia e Letteratura* 11 (1884), pp. 139–152.

Tounta 2008: Ε. Τούντα, *Το Δυτικό Sacrum Imperium και η Βυζαντινή Αυτοκρατορία: Ιδεολογικές τριβές και αλληλεπιδράσεις στην Ευρωπαϊκή πολιτική σκηνή του 12ου αιώνα (1135–1177)*. Αθήνα: Ίδρυμα της Ριζαρείου Εκκλησιαστικής Σχολής, 2008.

Tourptsoglou-Stefanidou 1998: Β. Τουρπτσόγλου-Στεφανίδου, *Περίγραμμα Βυζαντινών οικοδομικών περιορισμών. Από τον Ιουστινιανό στον Αρμενόπουλο και η προβολή τους στη νομοθεσία του Νεοελληνικού Κράτους*. Θεσσαλονίκη: Εταιρεία Μακεδονικών Σπουδών, 1998.

Trevisanato 2007: S. Trevisanato, "The Biblical plague of the Philistines now has a name: tularemia", *Medical Hypothesis* 69 (5) (2007), pp. 1144–1146.

Trivedi 2016: K. Trivedi, *Probability and Statistics with Reliability, Queuing, and Computer Science Applications*. Hoboken, NJ: John Wiley & Sons, 2016.

Tsiamis 2009: C. Tsiamis, E. Poulakou-Rebalakou, and E. Petridou, "The Red Sea and the port of Clysma: a possible gate of Justinian's plague", *Gesnerus* 66 (2) (2009), pp. 209–217.

Tsiamis 2013: C. Tsiamis et al., "Earthquakes and plague during Byzantine Times: Can lessons from the past improve epidemic preparedness?", *Acta Medico-Historica Adriatica* 11 (1) (2013), pp. 55–64.

Tsiamis 2014a: C. Tsiamis et al., "The Venetian Lazarettos of Candia and the Great Plague (1592–1595)", *Le infezioni in medicina. Rivista periodica di eziologia, epidemiologia, diagnostica, clinica e terapia delle patologie infettive* 22 (1) (2014), pp. 69–82.

Tsiamis 2014b: C. Tsiamis, E. Poulakou-Rebelakou, and G. Androutsos, "The role of the Egyptian Sea and land routes in Justinian plague: the case of Pelusium", in D. Michaelides (ed.), *Medicine and Healing in Ancient Mediterranean, including the Proceedings of the International Conference with the same title, organised in the framework of the Research Project 'INTERREG IIIAA: Greece-Cyprus 2000–2006, Joint Educational and Research Programmes in the History and Archaeology of Medicine, Palaeopathology and Palaeoradiation' and the 1st International CAPP Symposium 'New Approaches to Archaeological Human remains in Cyprus'*. Oxford, and Philadelphia, PA: Oxbow, 2014, pp. 334–337.

Tsiamis 2018a: C. Tsiamis et al., "Quarantine and British "protection" of the Ionian Islands, 1815–64", in J. Chircop, and F. Martinez (eds), *Mediterranean Quarantines, 1750–1914: Space, Identity and Power*. Manchester: Manchester University Press, 2018, pp. 256–279.

Tsiamis 2018b: C. Tsiamis et al., "The Knights Hospitaller of Rhodes and the Black Death of 1498: a poetic description of the plague", *Le infezioni in medicina. Rivista periodica di eziologia, epidemiologia, diagnostica, clinica e terapia delle patologie infettive* 26 (3) (2018), pp. 283–294.

Tsiros 2009: E. Tsiros, C. Domenikiotis, and N. Dalezios, "Sustainable production zoning for agroclimatic classification using GIS and remote sensing", *Journal of the Hungarian Meteorological Society (OMSZ)* 113 (1–2) (2009), pp. 55–68.

Turner 1997: D. Turner et al., "An Investigation of Polymorphism in the Interleukin-10 Gene Promoter", *European Journal of Immunogenetics* 24 (1) (1997), pp. 1–8.

Twigg 1978: G. Twigg, "The role of rodents in plague dissemination: a worldwide review". *Mammal Review* 8 (3) (1978), pp. 77–110.

US Department of Health and Human Services 2007: US Department of Health and Human Services, Centers for Disease Control and Prevention, and National Institutes of Health, *Biosafety in microbiological and biomedical laboratories*. Washington, D.C.: U.S. Government Printing Office, 2007.

Valiakos 2019: I. Valiakos, *Das Dynameron des Nikolaos Myrepsos. Erstedition*. Heidelberg: Propyleum, 2019.

Valpy 1828: A. Valpy, *An Etymological Dictionary of the Latin Language*. London: Baldwin-Longman-Whittaker, 1828.

Valtueña 2017: A. Valtueña et al., "The Stone Age Plague and its persistence in Eurasia", *Current Biology* 27 (2017), pp. 3683–3691.

Van Loghem 1944: J. Van Loghem, "The classification of plague bacillus", *Antoine van Leeuwenhoek Journal of Microbiology and Serolory* 10 (1994), pp. 15–16.

Varlik 2015: N. Varlik, *Plague and Empire in the Early Modern Mediterranean World. The Ottoman experience 1347–1600*. New York, NY: Cambridge University Press, 2015.

Velimirovich 1990: B. Velimirovich, "Plague and glasnost. First information about human cases in the USSR in 1989 and 1990", *Infection* 18 (1990), pp. 388–393.

Vlyssidou 1998: Β. Βλυσίδου, "Θέμα Θρακησίων", in E. Kountoura-Galake, St. Lampakes, T. Lounghis, A. Savvides, and V. Vlyssidou, *Asia Minor and its Themes. Studies on the Geography and Prosopography of the Byzantine Themes of Asia Minor (7th-11th century)* (The National Hellenic Research Foundation, Institute for Byzantine Research, Research Series 1). Athens: National Hellenic Research Foundation, Institute for Byzantine Research, 1998, pp. 201–234.

Vogler 2016: A. Vogler, P. Keim, and D. Wagner, "A review of methods of subtyping *Yersinia pestis*: From phenotypes to whole genome sequencing", *Infection Genetics and Evolution* 37 (2016), pp. 21–36.

Volk 1996: W. Volk, "*Yersinia*", in W. Volk et al. (eds), *Essentials of Medical Microbiology*. New York, NY: Lippincot-Raven, 1996, pp. 387–391.

von Kremer 1880: A. von Kremer, *Über die großen Seuchen des Orients nach arabischen Quellen* (Sitzungsberichte der kaiserlichen Akademie der Wissenschaften zu Wien, Philophish-Historischen Classe 96). Wien: Kaiserliche Akademie der Wissenschaften, 1880.

Wagner 2014: D. Wagner et al., "*Yersinia pestis* and the Plague of Justinian 541–543 AD: a genomic analysis", *Lancet Infectious Diseases* 14 (2014), pp. 319–326.

Walløe 2008: L. Walløe, "Medieval and Modern Bubonic Plague: Some Clinical Continuities", *Bulletin of the History of Medicine*, Supplement 27 (2008), pp. 59–73.

Ward 2007: W. Ward, "Aila and Clysma: The rise of northern ports in the Red Sea in Late Antiquity", in J. Starkey et al. (eds), *Natural Resources and Cultural Connections of the Red Sea* (BAR International Series 1661). Oxford: British Archaeological Reports, 2007, pp. 161–171.

Warner 1992: J. Warner, "Before there was "alcoholism": lessons from the medieval experience with alcohol", *Contemporary Drug Problems* 19 (1992), pp. 409–429.

Watson 2007: J. Watson, M. Gayer, and M. Connolly, "Epidemics after natural disasters", *Emerging Infectious Diseases* 13 (1) (2007), pp. 1–5.

Wayne 1986: L. Wayne, "Actions of the Judicial Commission of the International Committee on Systematic Bacteriology on requests for opinions published in 1983 and 1984", *International Journal of Systematic Bacteriology* 36 (1986), pp. 357–358.

Weil 2018: W. Weil et al., "Mutation landscape of base substitutions, duplications, and deletions in the representative current cholera pandemic strain", *Genome Biology and Evolution* 10 (8) (2018), pp. 2072–2085.

Weiss 2004: R. Weiss and A. McMichael, "Social and environmental risk factors in the emergence of infectious diseases", *Nature Medicine* 10 (12 Supplement) (2004), S70-S76.

Wenzel 1996: R. Wenzel et al., "Current understanding of sepsis", *Clinical Infectious Diseases* 2 (1996), pp. 401–413.

Wheelis 2002: M. Wheelis, "Biological Warfare at the 1346 Siege of Kaffa", *Emerging Infectious Diseases* 8 (2), pp. 971–975.

Whittam 1996: T. Whittam, "Variation and evolutionary processes in natural populations of *Escherichia coli*", in F. Neidhardt et al. (eds), *Escherichia coli and Salmonella*. Washington, D.C.: American Society for Microbiology, 1996, pp. 2707–2720.

Wiley 1998: T. Wiley et al., "Impact of alcohol on the histological and clinical progression of hepatitis C infection", *Hepatology* 28 (1998), pp. 805–809.

Wilkinson 1981: J. Wilkinson, *Egeria's Travels to the Holy Land*. Jerusalem: Ariel, 1981.

Willersley 2005: E. Willersley, and A. Cooper, "Ancient DNA", *Proceedings of the Royal Society*, B: *Biological Sciences* 272 (1558) (2005), pp. 3–16.

Willet 1992: H. Willet, "Physiology of Bacterial Growth", in W. Jocklik (ed.), *Zinsser's Microbiology*. Norwalk: Appleton & Lange, 1992, pp. 44–62.

Williams 1984: J. Williams, "Proposal to reject the new combination *Yersinia pseudotuberculosis* subsp. *pestis* for violation of the first principle of the International Code of Nomenclature of Bacteria Request for an opinion", *International Journal of Systematic Bacteriology* 34 (1984), pp. 268–269.

Williams 2000: K. Williams, "Investigation into the role of the serine protease HtrA in *Yersinia pestis* pathogenesis", *Federation of European Microbiological Societies (FEMS) Microbiology Letters* 176 (2) (2000), pp. 281–286.

Wispelwey 2008: B. Wispelwey, *Biographical Index of the Middle Ages*, volume 1: A-I München: K.-G. Saur, 2008.

Witakowski 1991: W. Witakowski, "Sources of Pseudo-Dionysius for the Third Part of his *Chronicle*", *Orientalia Suecana* 40 (1991), pp. 252–275.

World Health Organization 1950: World Health Organization, "Expert Committee on Plague. Report of the First Session" (WHO Technical Report Series 11). Geneva: World Health Organization, 1950, pp. 1–191.

World Health Organization 1968: *The work of WHO in the South-East Asia Region: Report of the Regional Director 1 July 1967–1 August 1968 (SEA/RC21/2)*. New Delhi: World Health Organization, Regional Office for South-East Asia, 1968.

World Health Organization 1996: World Health Organization, "Human Plague in 1994", *WHO Weekly Epidemiological Record* 122 (1996), pp. 165–168.

World Health Organization 2003: World Health Organization, "Human Plague in 2000 and 2001", *WHO Weekly Epidemiological Record* 78 (16) (2003), pp. 130–136.

World Health Organization 2015: *International Classification Diseases for Mortality and Morbidity statistics*. 11th Revision. Geneva: World Health Organization, 2015.

Wu 1936: L. Wu et al., *Plague: A Manual for Medical and Public Health Workers*. Shanghai: Weishengshu National Quarantine Service, Shanghai Station, 1936.

Wunderrlich 2018: J. Wunderrlich, R. Acuña-Soto, and W. Alonso, "Dengue hospitalizations in Brazil: annual wave from West to East and recent increase among children", *Epidemiology and Infection* 146 (2) (2018), pp. 236–245.

Xia 1982: W. Xia et al., "Population Dynamics and Regulation of Mongolia Gerbils (*Meriones unguiculatus*) in North Yinshan Mountain, Inner Mongolia of China", *Acta Theriologica Sinica* 2 (1982), pp. 51–71.

Xua 2011: L. Xua et al., "Nonlinear effect of climate on plague during the third pandemic in China", *Proceedings of the National Academy of Science of the United States of America* 108 (25) (2011), pp. 10214–10219.

Young 2001: G. Young, *Rome's Eastern Trade: International Commerce and Imperial Policy 31 B.C.– 305 A.D.* London: Routledge, 2001.

Yue 2018: R. Yue, and H. Lee, "Climate change and plague history in Europe", *Science China Earth Sciences* 61 (2) (2018), pp. 163–177.

Yujing 2016: B. Yujing, "Immunology of *Yersinia pestis* infection", *Advances in Experimental Medicine and Biology* 918 (2016), pp. 273–293.

Zavialov 2003: A. Zavialov et al., "Stucture and biogenesis of the capsular F1 antigen form the *Yersinia pestis:* preserved folding energy drives fiber formation", *Cell* 113 (2003), pp. 587–596.

Zhang 2007: Z. Zhang et al., "Relationship between increase rate of human plague in China and global climate index as revealed by cross-spectral and cross-wavelet analyses", *International Journal of Zoology* 2 (2007), pp. 144–153.

Zhang 2008: P. Zhang et al., "Alcohol abuse, immunosuppression, and pulmonary infection", *Current Drug Abuse Reviews* 1 (2008), pp. 56–67.

Zhang 2014: R. Zhang et al., "Identification of Horizontally-transferred Genomic Islands and Genome Segmentation Points by Using the GC Profile Method", *Current Genomics* 15 (2014), pp. 113–121.

Zhou 2004: D. Zhou et al., "Genetics of Metabolic Variations between *Yersinia pestis* Biovars and the Proposal of a new Biovar: Microtus", *Journal of Bacteriology* 186 (15) (2004), pp. 5147–5152.

Zhou 2006: D. Zhou, Y. Han, and R. Yang, "Molecular and physiological insights into plague transmission, virulence and etiology", *Microbes and Infections* 8 (2006), pp. 273–284.

Ziegler 2003: P. Ziegler, *The Black Death.* London: Sutton, 2003.

Zimmerman 1992: M. Zimmerman, "Die lykischen Häfen und die Handelswege im östlichen Mittelmeer", *Zeitschrift für Papyrologie und Epirgraphik* 92 (1992), pp. 215–216.

Zykin 1994: L. Zykin, "The results of a 15-year study of the persistence of *Yersinia pestis* L-forms", *Journal of Microbiology, Epidemiology, and Immunobiology* Supplement 1 (1994), pp. 68–71.

Glossary

Aerobic bacteria: Bacteria which require oxygen for their growth, reproduction, and energy needs.

Adaptive immunity: A type of immunity. Immune response of the human organism to the invasion of specific agents (antigens).

Agar: Growth media for microorganisms.

Aminoacids: Organic compounds that contain an amine group, a carboxylic acid group, and a side-chain. Amino acids are basic units of proteins.

Anaerobic bacteria: Bacteria which do not require oxygen for their growth, reproduction, and energy needs.

Anaerobiosis: The life of bacteria in conditions of absence of oxygen.

Anthrax: Infectious bacterial zoonotic disease caused by *Bacillus anthracis*. Transmission to man by contact with infected animals. *Bacillus anthracis* is classified as Category A Biological weapon.

Apoptosis: The process of programmed cell death.

ARDS syndrome: The Acute Respiratory Distress Syndrome (ARDS) is a rapid, progressive disease in critically ill patients. It is characterized by pulmonary infiltration, severe hypoxemia, and short-ness of breath.

Artificial immunity: Type of (adaptive) immunity induced by vaccines (active artificial immunity) or injection of antibodies into non-immune persons (passive artificial immunity).

Bacteraemia: The invention of infectious agents in the blood stream.

Bacteriophage: Viruses that infect and replicate themselves inside bacteria.

Bacteriocins: Bactericial antibiotic-like substance with killing action against same or closely related species.

Basic Reproduction Number: The number of new secondary cases (infected persons) generated by a single case (infected person).

Binary fission: Process of asexual reproduction with a simple cellular division.

Biovars: Strains with different biochemical properties (See Strains).

Bubo: Inflamed lymph node.

Burrow index: Average number of free living fleas per species per rodent burrow.

Carbuncle: Necrotizing skin infection in the formation of abscess usually due to *Staphylococcus aureus*.

Capsule: Basic bacterial structure as a layer outside the cell wall. The capsule protects bacteria from phagocytosis, desiccation complement-mediated lysis and the oxygen toxicity. The capsule is a major virulent factor.

Cat-scratch disease: Infectious disease that causes the swelling of the lymph node, near to the area of the infected skin, after a cat scratch or bite. The disease is caused by *Bartonella henselae*.

Cheopsis index: Average number of *Xenopsylla cheopis* per rat.

Cowpox: Mild, self-limited skin disease of cows caused by cowpox virus (resembling mild smallpox). Human can be infected due to direct contact with the infected animal.

Cyanosis: Bluish discoloration of the skin due to a high concentration of reduced hemoglobin in the blood.

Complement system: Part of the immune system. The activation of this complex of non-specific pro-teins (more than 20) leads to changes on the surface of the pathogen, and its destruction.

DIC syndrome: The Disseminated Intravascular Coagulation (DIC) (coagulopathy) syndrome is charac-terized by intravascular over-activation of coagulation. DIC leads to microvascular dysfunction and organ failure.

https://doi.org/10.1515/9783110613636-016

DNA: Nucleic acid that contains the genetic information for the development of the living organisms. DNA is found in the double helix, and its basic units are the nucleotides. Each nucleotide contains a sugar (deoxyribose), a phosphate group, and a nitrogenus base. The nucleotides in the double helix are distributed in pairs: adenine/thymine, cytosine/guanine.

DNA polymerase: Type of enzyme that creates new copies of DNA molecules in the form of nucleic acids.

Ecchymosis: Discoloration of the skin caused by the collection of blood from ruptured vascular capillaries into subcutaneous tissues.

Enzyme: Protein that accelerates a chemical reaction without being altered during the reaction itself.

Ergotism: Poisoning due to the consumption of ergot-infected grain. Ergot is a fungus (*Claviceps purpurea*) that infects cereals.

Ethology: Study of animal behavior in their natural environmental.

Eukariotic cell: Type of living cells of higher animals and plants, containing the nucleus bounded by the nuclear membrane, and organelles with cellular functions.

Facultative anaerobe: Organism that can survive in the presence of oxygen (aerobic respiration), or in anaerobic conditions via fermentation or anaerobic respiration.

Filariasis: Parasitic disease caused by filarial worms (*Wuchereria bancrofti* in 90% of the cases and *Brugia malayi*). The disease is endemic in tropical and subtropical zones, and it is transmitted to humans by the bite of infected mosquitoes. The adult forms of the worms reside in the lymphatic system leading to obstruction of the lymph flow.

Gram-negative bacterium: Classification of bacteria based on Gram's staining method (losing the stain or decolorized by alcohol). The cell wall of the gram-negative bacteria contains a thin layer of peptidoglycan covered by a membrane of lipopolysaccharide and lipoprotein. Under the microscope, they are recognized as pink formations.

Gram-positive bacterium: Classification of bacteria based on Gram's staining method (retaining the stain or resisting by alcohol). The cell wall of the gram-positive bacteria contains a thick layer of peptidoglycan with attached teichoic acids. Under the microscope, they are recognized as violet formations.

Horizontal gene transfer (HGT): Process of genetic material incorporation in the genome of an organism from the genome of another organism.

Incubation period: The period of time between the entrance of a microorganism in the organism and the onset of the symptoms.

Innate immunity: Phylogenically the oldest type of immunity. It is the response to the pathogen invasion due to the anatomical and physiological barriers, inflammation, complement activation, cytokine secretion, target cell lysis, and phagocytosis.

Insertion sequences (IS) elements: The simplest type of transposable genetic elements. They are small fragments of DNA which move within or between genomes carrying new characteristics in the bacteria (i.e. antibiotic resistance).

L-forms bacteria: Bacterial variants as a result of the complete or partial inhibition of cell wall synthesis.

Lassa hemorrhagic fever: Acute viral haemorrhagic disease endemic to West Africa. The disease is caused by Lassa Virus. It is transmitted from rodents (mastomys species) to humans, and also from human to human due to fluid exchange or in hospitals (reused needles, or contaminated equipment, for example). Symptoms include fever, headache, facial swelling, fatigue, nausea, vomiting, diarrhea, conjunctivitis, and mucosal bleeding.

Legionnaire's Disease: Type of pneumonia caused by *Legionella pneumophilia* which grows in warm water (cooling towers, hot tubs, showers). The symptoms include high fever, headache, cough, and dyspnea.

Leptospirosis: Infectious disease of humans and animals (rats, wild rodents, dogs, swine, cattle) caused by bacteria of genus Leptospira. The symptoms include mostly high fever, abdominal pain, vomiting, diarrhea, and dehydration. Leptospira bacteria are found worldwide, but mainly in tropical zones. The disease is transmitted to humans by direct exposure with urine or tissue of infected animals.

Logarithmic growth (bacterial): One of the four phases of the bacterial growth (initial, exponential, stationary, death). During the Log phase (exponential phase) the bacterial cells are dividing and doubling in numbers after a generation time.

Lymphogranuloma venereum: Acute and chronic sexually transmitted disease caused by *Chlamydia trachomatis*.

Macrophages: A type of white blood cells. They are produced by the differentiation of monocytes in tissue.

Meiosis: A type of cell division. The result of meiosis is four new cells with a reduction of the number of chromosomes by half.

Microbiota (Human): Collection of specific microorganisms (bacteria, fungi, viruses, archaea) in different parts of the body (oral, gastrointestinal, vaginal, or skin). The term "microbiome" refers to the genome of these microorganisms.

Micro-evolution: Evolutionary change over a short period of time leading to new varieties within a species.

Mitosis: A type of cell division in eukaryotes. The cell is divided into two daughter cells, genetically identical.

mRNA: The messenger RNA is a single stranded RNA. It carries genetic information from the nucleus to the ribosomes for the synthesis of the proteins. In the mRNA, a three-nucleotide sequence is called *codon*, and each codon corresponds to a particular amino acid.

Natural Killer Cells: A type of lymphocytes and component of the innate immunity.

Pathogenicity: The ability of an organism to cause disease.

Pathogenicity islands (PAIs): Clusters of genes with a vital role in the virulence of bacteria. The PAIs is acquired by bacteria through horizontal gene transfer.

Petechiae: Small red or purple spots (1–2 mm) caused by minor bleeding in the skin.

Plasmid: Extra-chromosomal ring-shaped molecule of DNA. Plasmids can be replicated independently of the bacterial chromosome.

Prokaryotes: Usually unicellular organisms. They belong to Bacteria and Archaea, two of the major domains of life. Contrary to the eukaryotes, they lack a distinct nucleus and other cellular organelles. Their genetic material forms a single continuous strand (or loops).

pH: The concentration of hydronium ions (H_3O^+) in a solution. As the concentration of ions decreases, the pH is higher (basic). As the concentration of ions increases, the pH is lower (acidic).

Phagocytosis: The cellular process of engulfing and destruction of solid particles and bacteria by the cell membrane.

Phenotype: The observable characteristics of the organisms (behaviour, biochemical properties, colour, shape, and size). They are determined by the genotype (set of genes) and environmental influences.

Phylogeny: The description of the evolutionary history of relationship between organisms.

Polytomy: Simultaneous genetic divergence of multiple lineage branches.

RNA: Ribonucleic Acid (RNA) is very similar to DNA, except that it is single-stranded. Also, RNA has the nitrogenus base uracil, instead of thymine. Uracil pairs with adenine. RNA transfers genetic information from nucleus to cell, i. e. from DNA to the proteins. RNA makes up ribosomes and proteins.
RNA polymerase: The enzyme capable to produce RNA from a DNA template.
Ribosome: Molecular structure in the cytosole of the eukaryotes and prokaryotes. It is the place for the production of proteins (See mRNA and RNA).
Ribotype: The type of ribosomal RNA in an organism.
Ribotyping: Method of genomic DNA identification based on differences in ribosomal RNA genes. The technique allows the identification of different strains of bacteria.

Septicemia (sepsis): Presence of bacteria in the bloodstream due to a rapidly progressive infection leading to multi-organ failure.
Serotherapy: Treatment of a disease by the injection of serum containing antibodies against a bacterium.
Serovars: Strains with different antigenic properties (See Strains).
Strain (bacterial): Group of microorganisms as genetic variant, or sub-type of a bacterium with particular characteristics. The term is used also to indicate a population of microbes descended from a single individual or a pure culture.

Taq polymerase (Taq): Thermostable enzyme used in PCR. Taq is isolated from bacterial genus *Thermus aquaticus*, and its most important feature is that it remains stable in high temperatures (100° C).
Thymus: Primary lymphoid organ behind sternum. It stimulates the development of T-cells.
Thyphoid fever: Infectious disease caused by *Salmonella typhi*. The disease is also known as enteric fever. Its main symptoms are high fever and abdominal pain/cramps. It is spread via faecal oral contamination. Typhoid fever is different disease from typhus (See Typhus).
Total flea index: Average number of fleas of all species per rat.
Tumor necrosis factor (TNF): Mediator (signaling protein) of the immune response and inflammatory reactions.
Tularemia: Infectious disease of the skin, eyes, lungs, and lymph nodes. It is caused by *Francisella tularensis*. The disease affects rodents, rabbits, hares, dogs, cats, sheep, and birds. It spreads to humans through insect bites (tick) or direct exposure to infected animals.
Typhus: Infectious disease spread by lice or fleas. Typhus is caused by *Rickettsia typhi*. Its main symptoms are high fever with chills and rigor, jaundice, abdominal pain, and vomiting. Typhus is different from thyphoid fever (See Typhoid fever).

Index of Names and Places

https://doi.org/10.1515/9783110613636-017

Subject Index

https://doi.org/10.1515/9783110613636-018